T0077719

FLORA OF GAUTAM BUDDHA WILDLIFE SANCTUARY, HAZARIBAG, JHARKHAND

(INDIA)

Acknowledgement

I am obliged to Principal Chief Conservator of Forests, Wildlife Division, Jharkhand and Mr. Kumar Manish Arvind, present Divisional Forest Officer (D.F.O.), Wildlife Division, Hazaribag for permitting me to do research in Gautam Buddha Wildlife Sanctuary, Hazaribag. I am also thankful to Dr. A.K. Mishra, D.F.O. Wildlife Division for his kind co-operation in my research work. Thanks are also due to Mr. Nalini Ranjan Prasad, Forester, Chouparan, Wildlife Division, Hazaribag and the Forest Guards of Wildlife for their support in my research work.

I am also thankful to Dr. V.J. Nair, Emeritus Scientist and former Director of Botanical Survey of India (B.S.I.), Southern Circle, Coimbatore for his kind encouragement and support in my research work. Thanks are also due to Director of B.S.I., Howrah to allow me to consult Central National Herbarium (C.N.H.) and Library. I am obliged to Dr. Lakshminarasimhan, Scientist-E and Joint Director of B.S.I., Howrah; Dr. R.K. Gupta, Scientist-C, B.S.I., Howrah; Dr. P.V. Sreekumar, Scientist-C of Aacharya Jagadish Chandra Bose Indian Botanic Garden, Howrah; Dr. P.R. Sur, Botanist, C.N.H.; Dr. Subir Bandyopadhyay, Botanist; Late Dr. M.K. Pathak, Botanist and Dr. D.K. Singh, Scientist-C; Mr. Shyam Biswa, Preservation Assistant Grade-I; Mr. Kanai Lal Maity, Curator, C.N.H.; Mr. Ajay Kumar Ghosh, Former Botanist, C.N.H.; Mr. Than Singh Niranjan, Senior Library Information Assistant; Mr. Dinesh Saha, Artist and all the officials and personnel of Botanical Survey of India, Howrah for their kind encouragement and support in my research work.

I am thankful to Dr. A.K. Pandey, Professor, Department of Botany, University of Delhi for giving idea of using herbaria in establishing phylogeny.

Thanks are also due to Dr. M.P. Sharma, Professor, Department of Botany, Hamdard University, Hamdard Nagar, New Delhi for giving the knowledge of Nomenclature and kind encouragement.

I am very thankful to Dr. N.D. Paria, Professor, Department of Botany, Ballygunj Science College, Calcutta University for taking interest in my research work and to encourage me.

I am obliged to Dr. H.B. Singh, Present Scientist G and Head, RHMD, National Institute of Science Communication And Information Resources (NISCAIR), New Delhi and Dr. M.V. Viswanathan, former scientist of NISCAIR for giving me techniques about Herbarium preparation.

I am thankful to Dr. Subhash Singh, Director and Mr. Vinod Kumar, Meteorological Observer, Soil Conservation Research and Training Centre, Demotand, Hazaribag, Jharkhand for giving me data of temperature, humidity and rainfall.

I am also thankful to Ms. Akanksha Lala, B.Sc. student, Kulti College, Kulti, West Bengal for her kind co-operation in making illustrations.

I am obliged to Dr. C. T. N. Singh who Guided me during my research work.

I am thankful to my parents especially my mother, Mrs. Madhuri Bala who is Retd. Teacher of Government Girls Middle School, Chouparan for her eternal support to my research work.

(Nirbhay Ambasta)

List of Plates Photographs and Illustrations

List of Tables

List of Graphs and Line Drawing

Dedicated to
my mother
Mrs. Madhuri Bala

PLATE - I

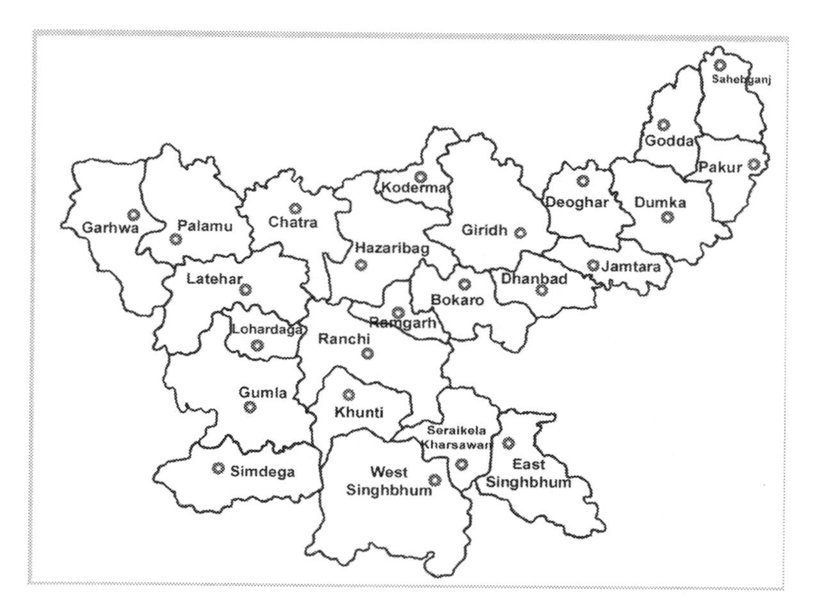

Fig.-1: Jharkhand State Map

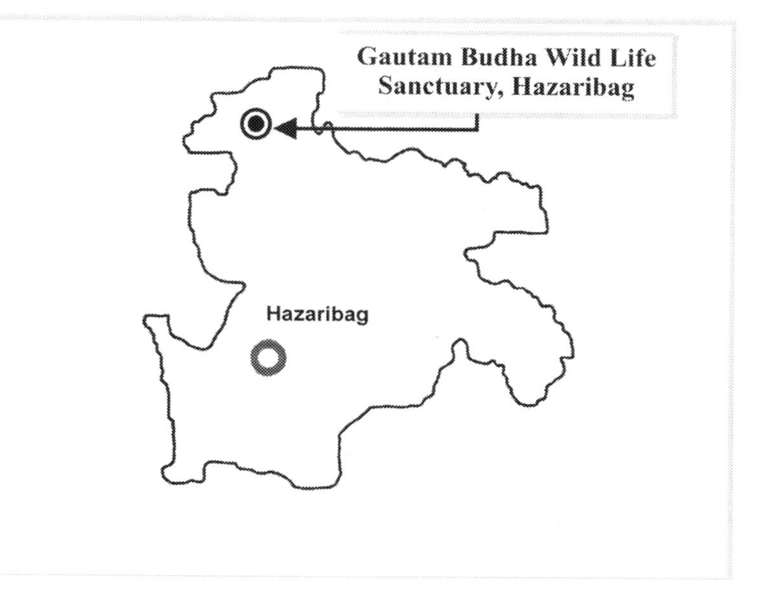

Fig.-2: Hazaribag District Map

PLATE - II

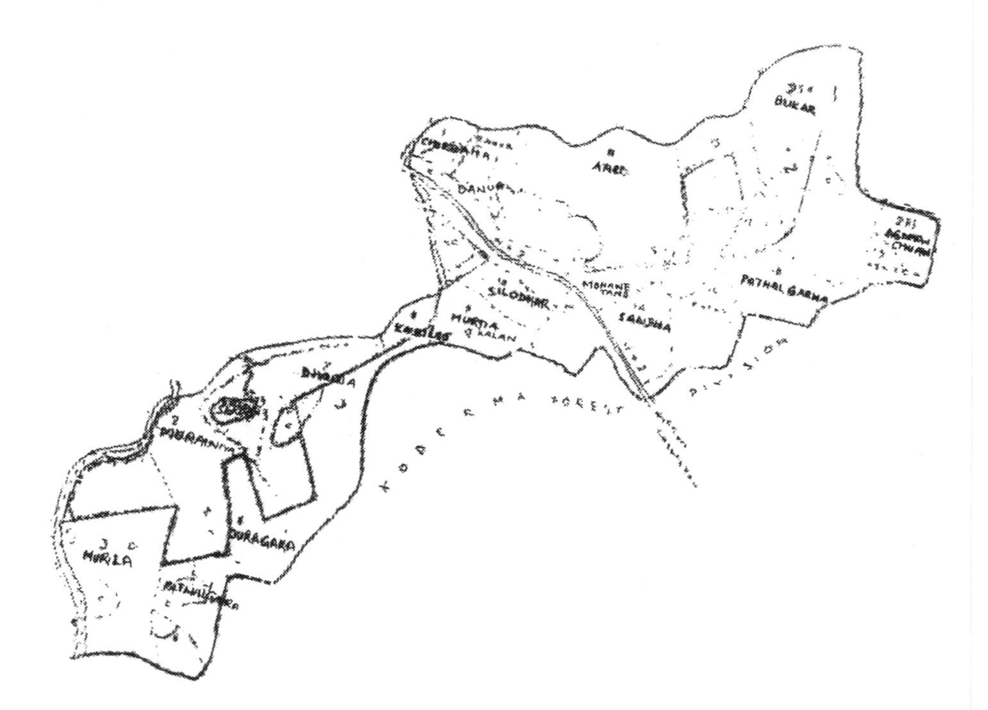

Fig.-3: Map of Gautam Buddha Wildlife Sanctuary, Hazaribag

PLATE - III

Fig.-4: NH-2 passing through Gautam Buddha Wildlife Sanctuary, Hazaribag

Fig.-5: Scenery of Gautam Buddha Wildlife Sanctuary, Hazaribag during rain

PLATE - IV

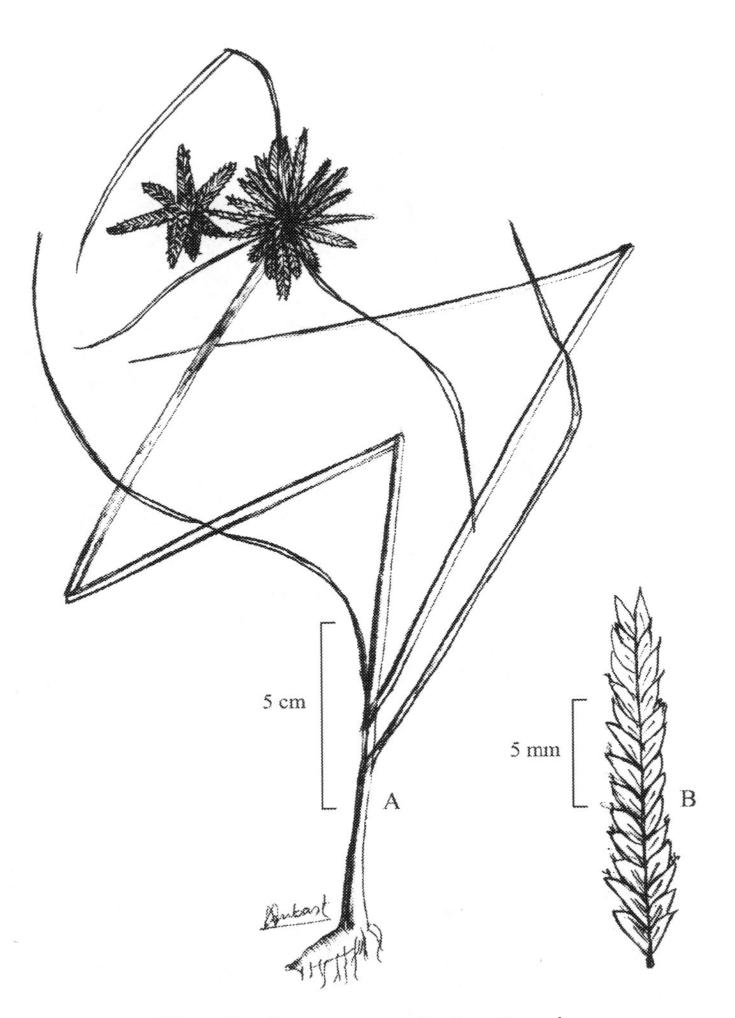

Fig.-6: *Cyperus tenuispica* Steud.

A. Plant
B. Spikelets

PLATE - V

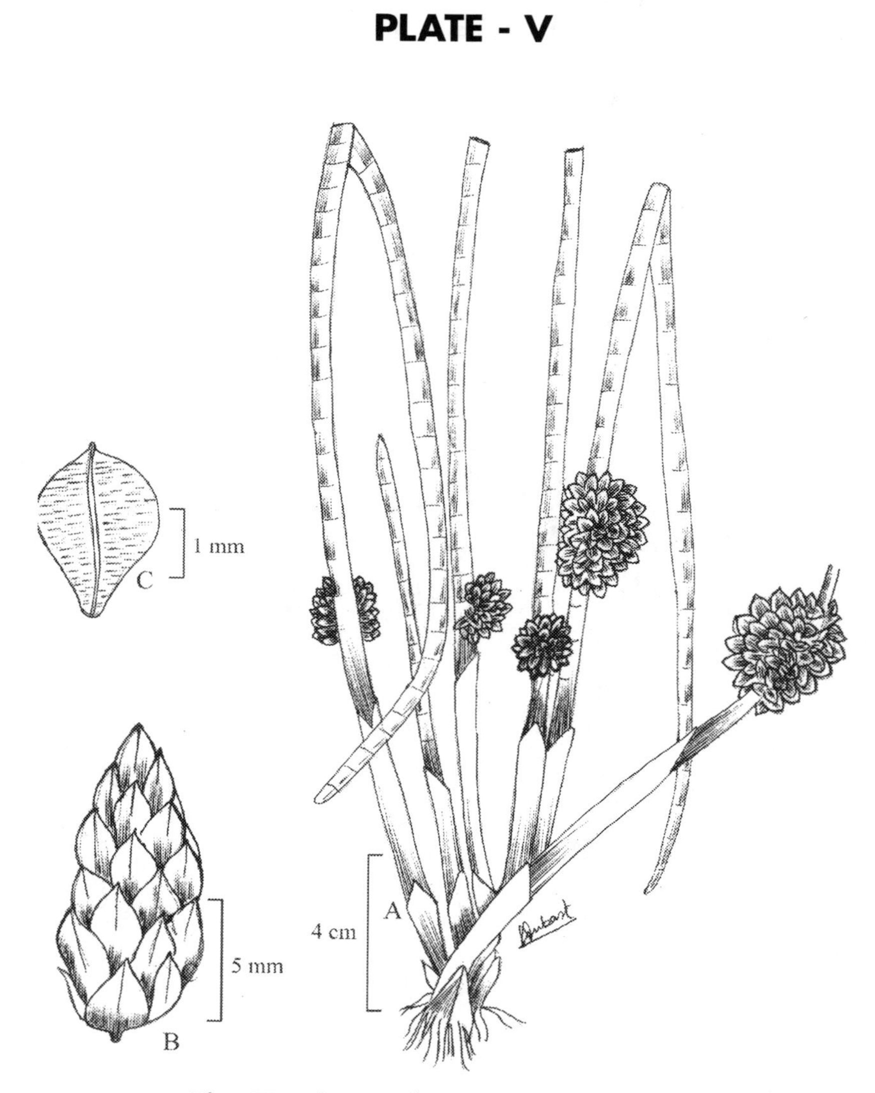

Fig.-7: *Schoenoplectus articulatus* (L.) Palla

A. Plant
B. Spikelet
C. Glume

PLATE - VI

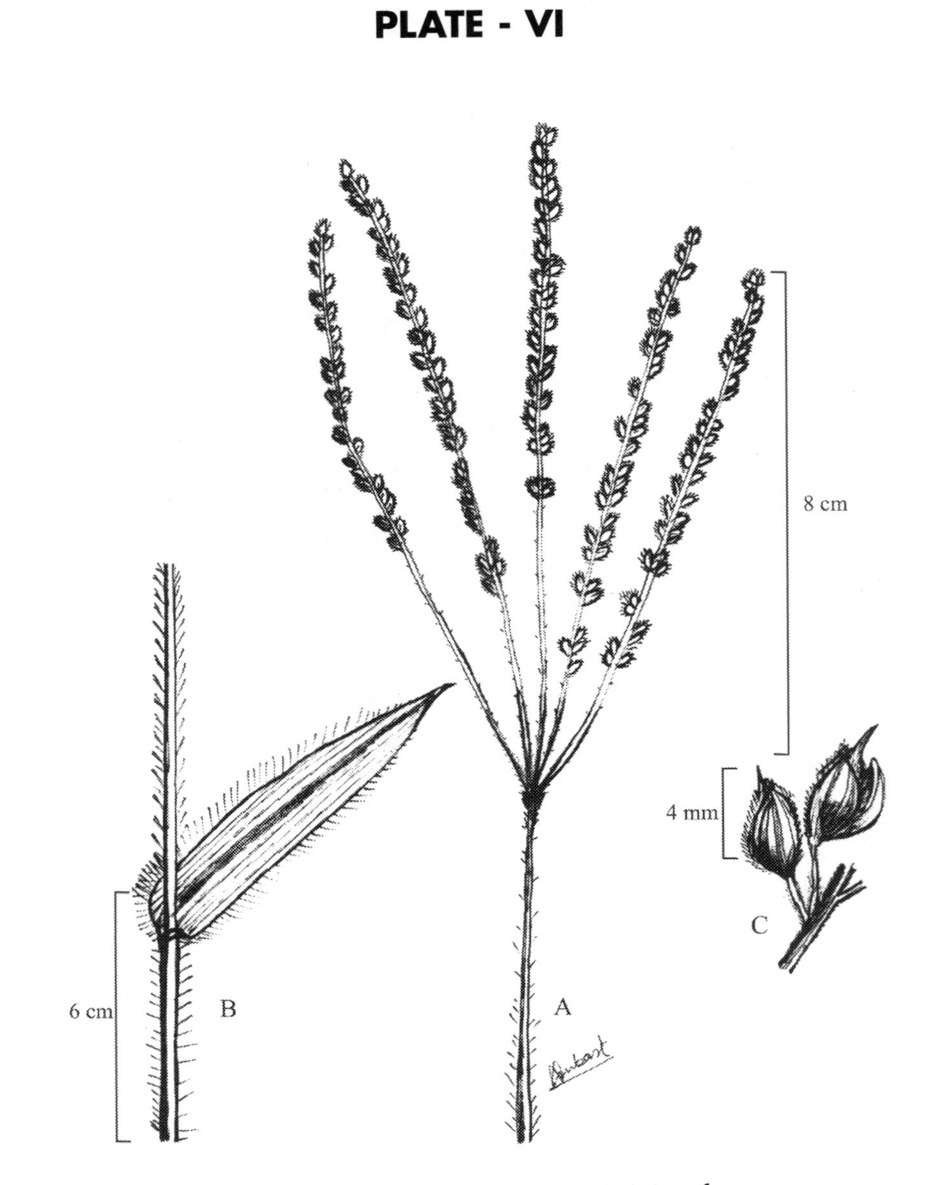

Fig.-8: *Alloteropsis cimicina* (L.) Stapf

A. Plant
B. Leaf, pectinately ciliate on the margins; stem hairy
C. Spikelets; margins bearded

PLATE - VII

Fig.-9: *Apluda mutica* L.

A. and B. Inflorescence and part of the geniculate base
C. A group of racemes each in its own special sheath
D. A group of spikelets. Note below the thick callus, on the left the awned fertile sessile spikelet and on the right two pedicels one of which carries a rudimentary, the other a perfect, spikelet

PLATE - VIII

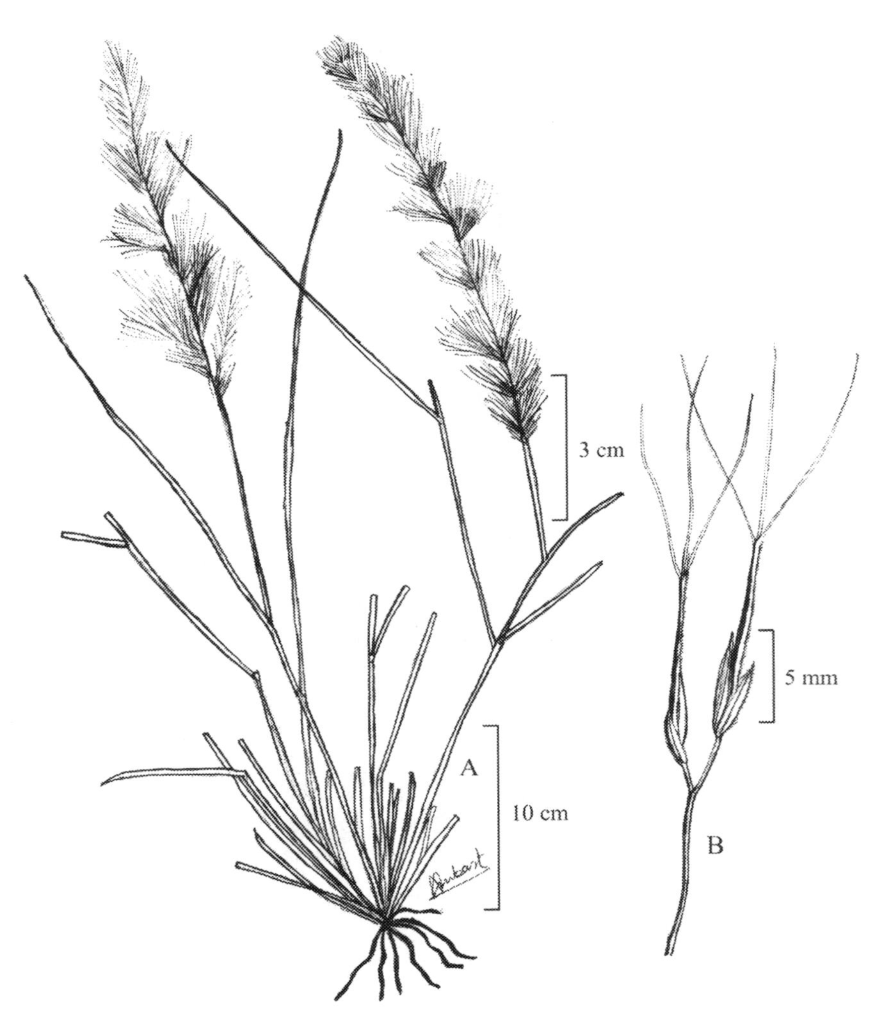

Fig.-10: *Aristida adscensionis* L.

A. Habit
B. Two spikelets showing the tripartite awn

PLATE - IX

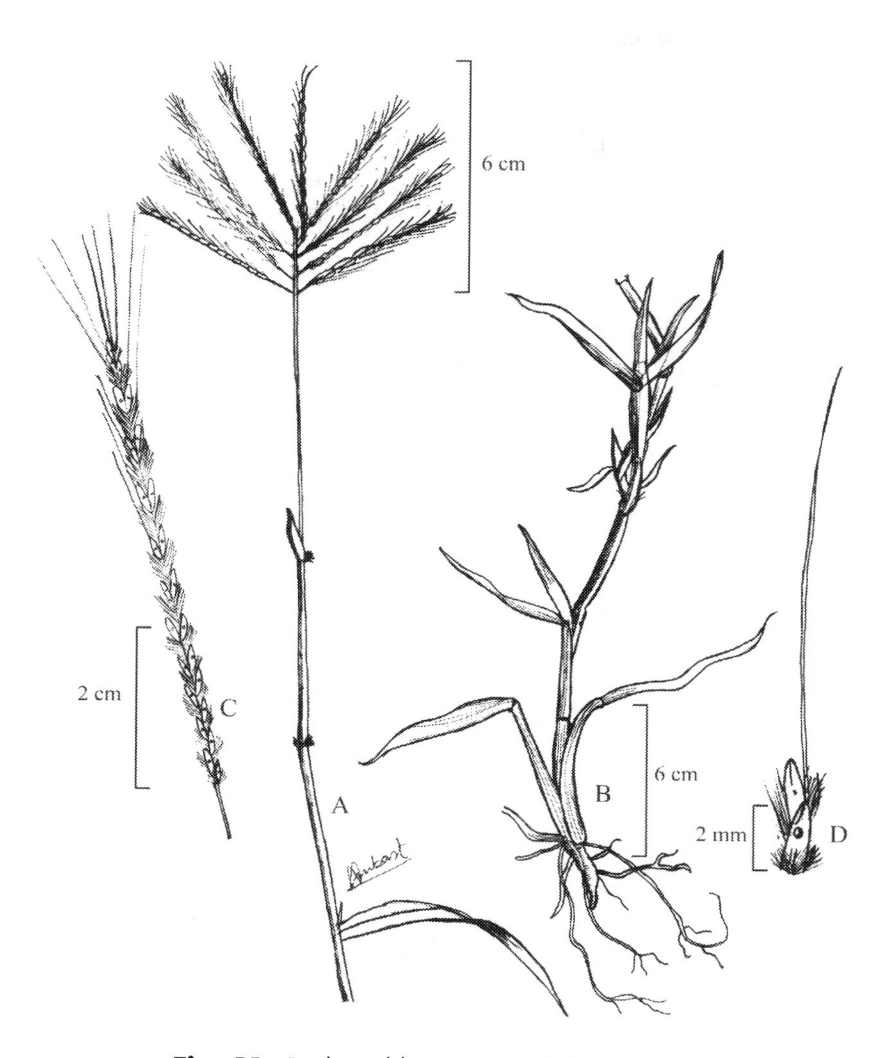

Fig.-11: *Bothriochloa pertusa* (L.) A. Camus

A. Inflorescence
B. Base of plant: a perennial
C. A mature raceme; the lower spikelets have shed their seeds and awns
D. Pair of spikelets, the lower awned; note the pit in the lower glume

PLATE - X

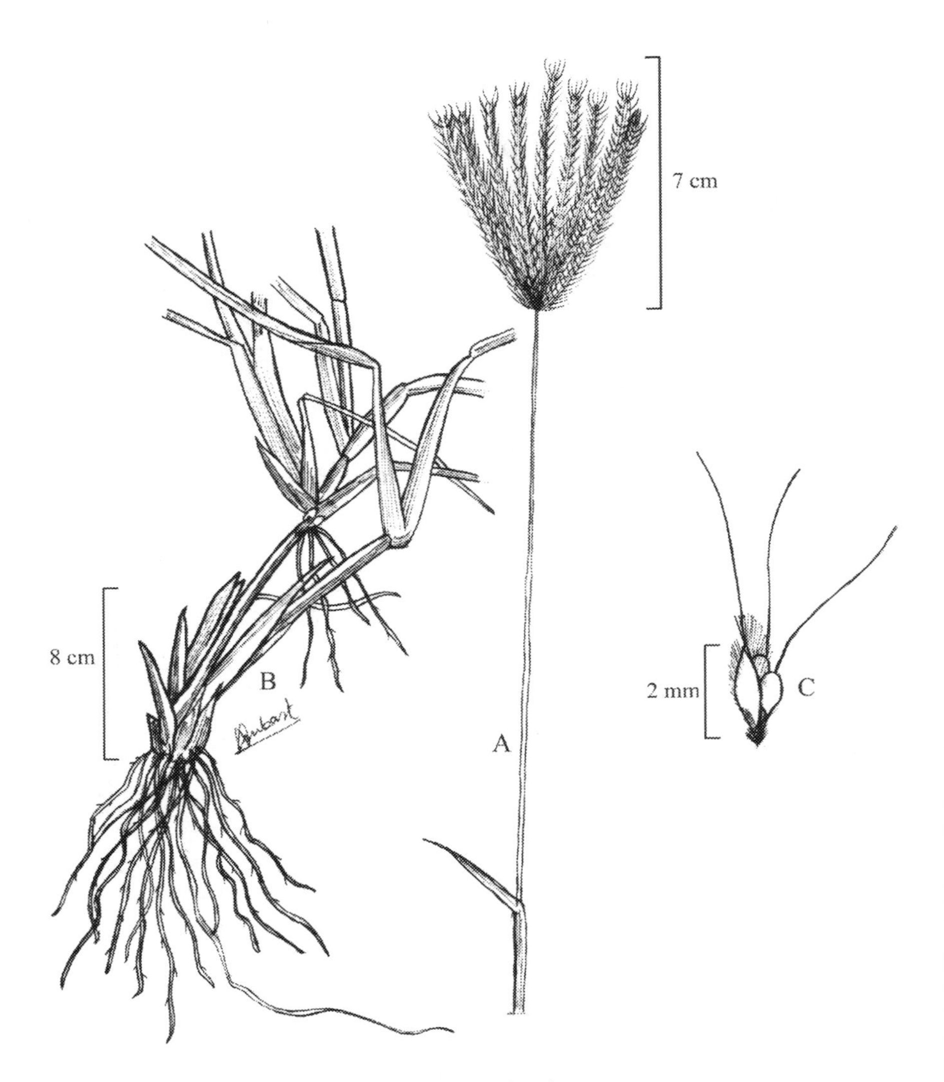

Fig.-12: *Chloris barbata* Sw.

A. Inflorescence of digitate ascending racemes
B. Base of plant: a perennial
C. Spikelet showing the awns

PLATE - XI

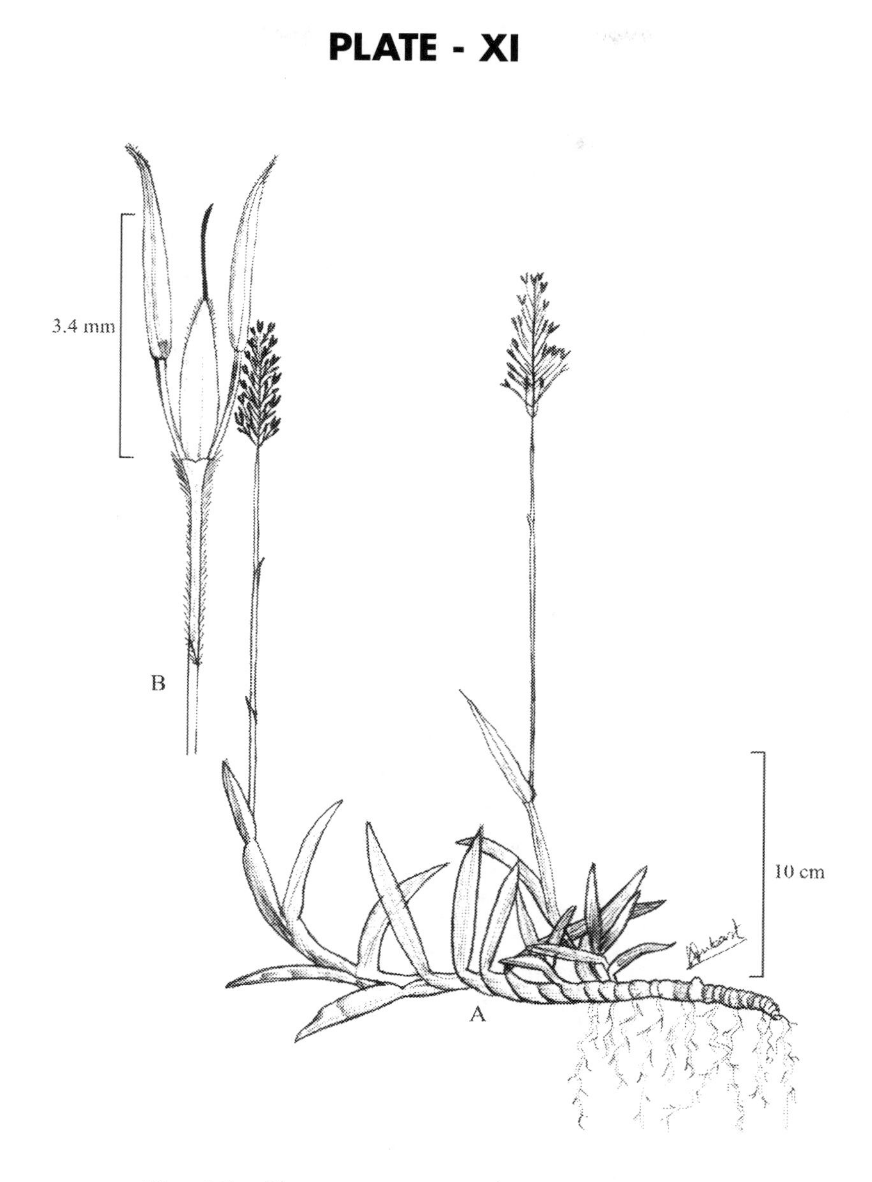

Fig.-13: *Chrysopogon aciculatus* (Retz.) Trin.

A. Whole plant showing its creeping habit, majority of leaves at the base and erect flowering culms

B. The triad of spikelets; note the sharp callus lying against the peduncle

PLATE - XII

Fig.-14: *Coix lachryma-jobi* L.

A. Part of the plant, showing the flowers

PLATE - XIII

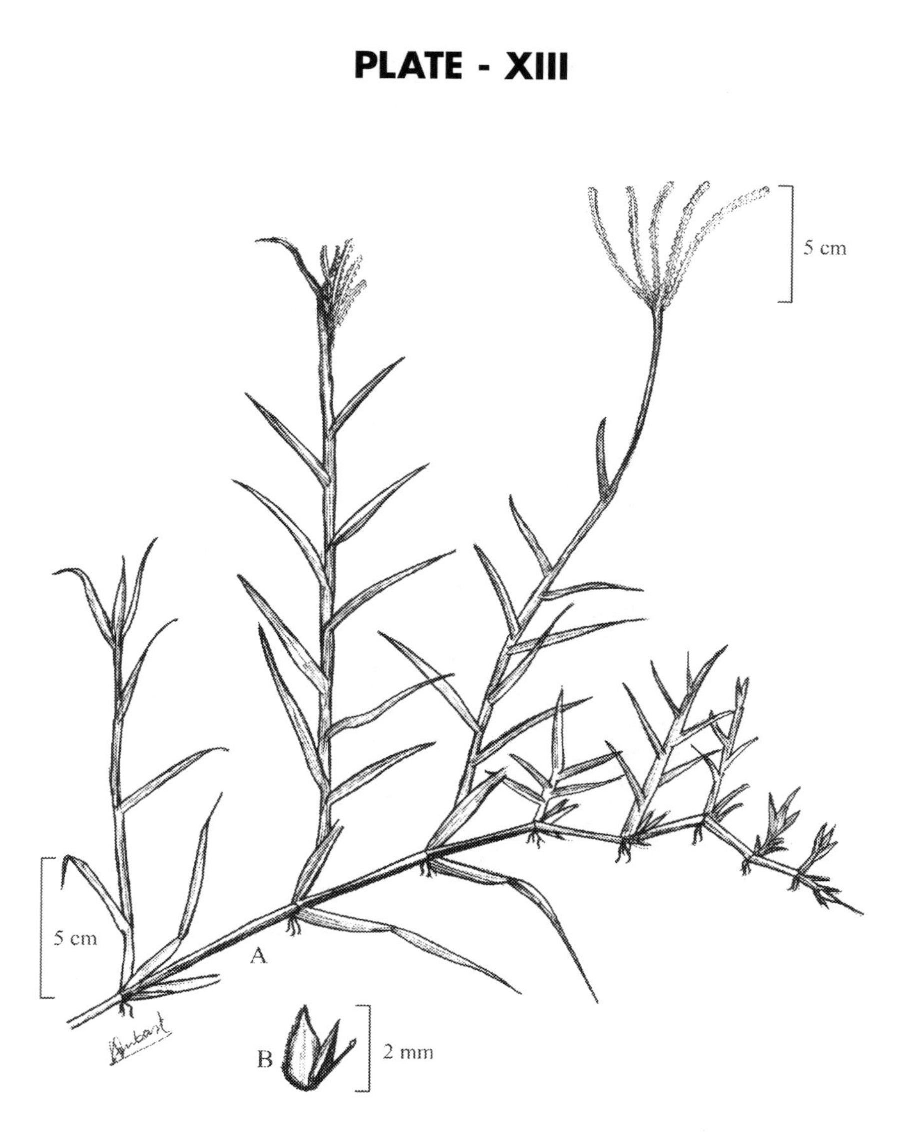

Fig.-15: *Cynodon dactylon* (L.) Pers.

A. Plant; note prostrate stem rooting at the nodes
B. A spikelet with glumes removed, showing the boat-shaped lemma and palea and the bristle-like produced rachilla

PLATE - XIV

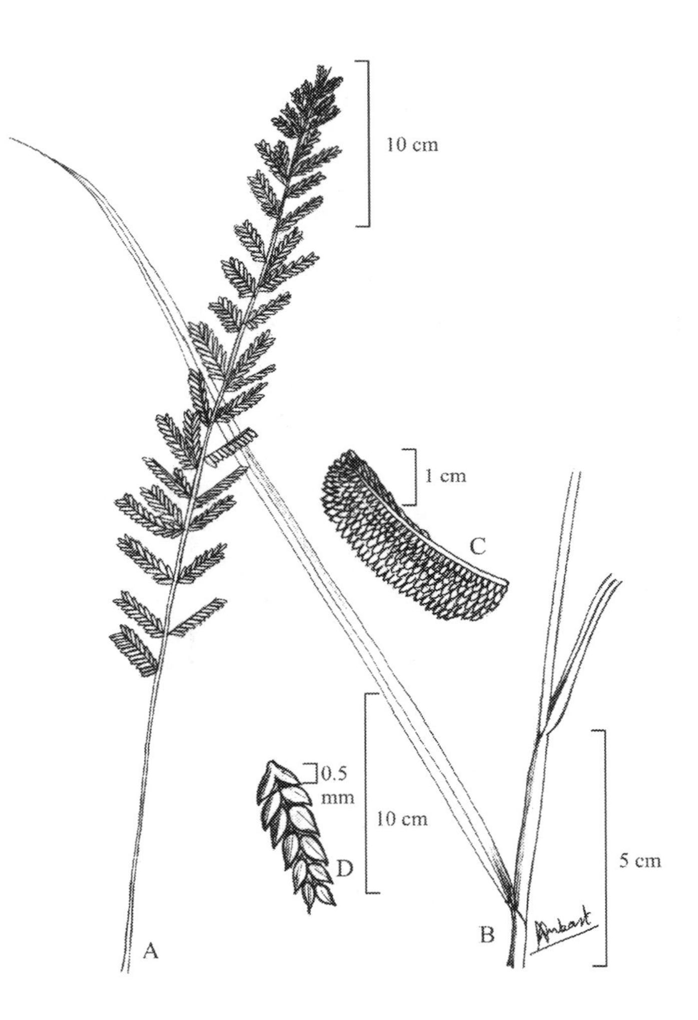

Fig.-16: *Desmostachya bipinnata* (L.) Stapf

A. Inflorescence of spreading closely spikeletted racemes
B. Stem
C. An individual raceme
D. A spikelet

PLATE - XV

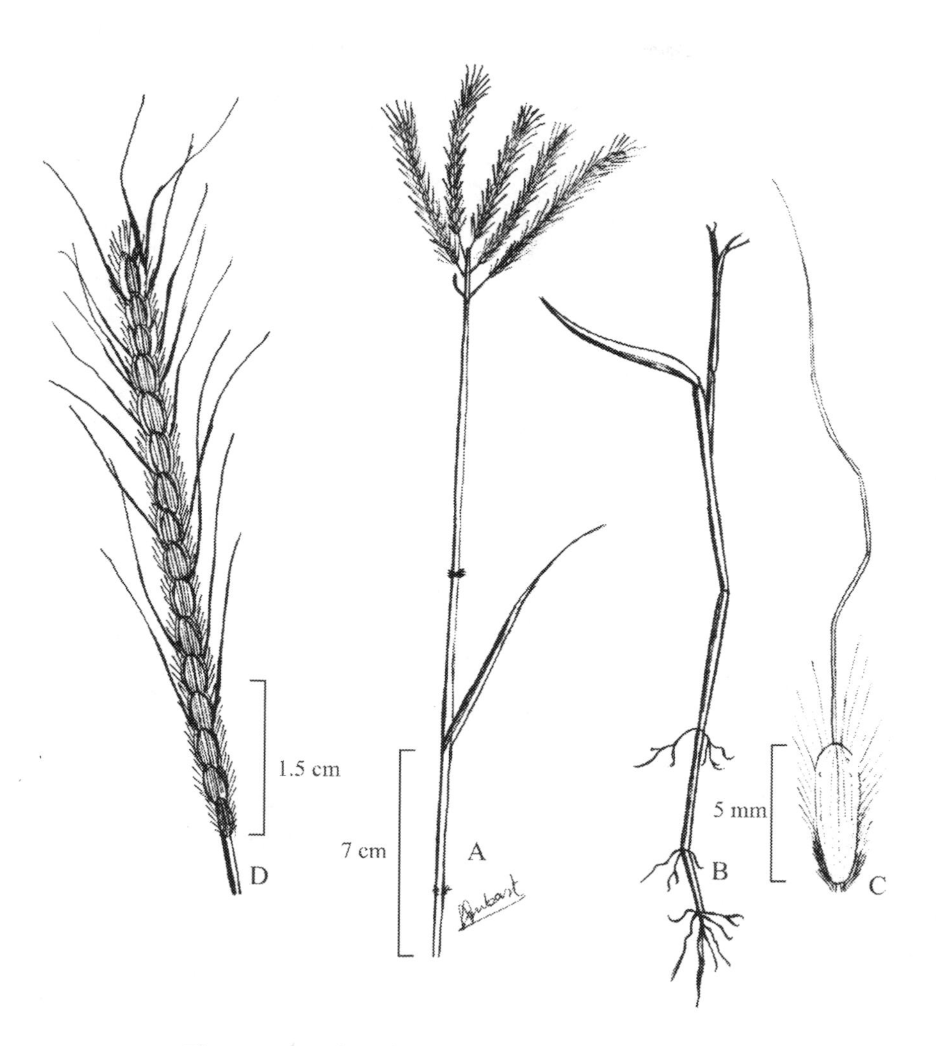

Fig.-17: *Dichanthium annulatum* (Forsk.) Stapf

A. Plant; note the bearded nodes
B. Creeping base, rooting at the nodes
C. Sessile spikelets; note the shape of the lower glume and the tubercle-based hairs on the margin
D. The manner in which the spikelets overlap is characteristic of the genus. The lower 2 or 3 pairs have no awns

PLATE - XVI

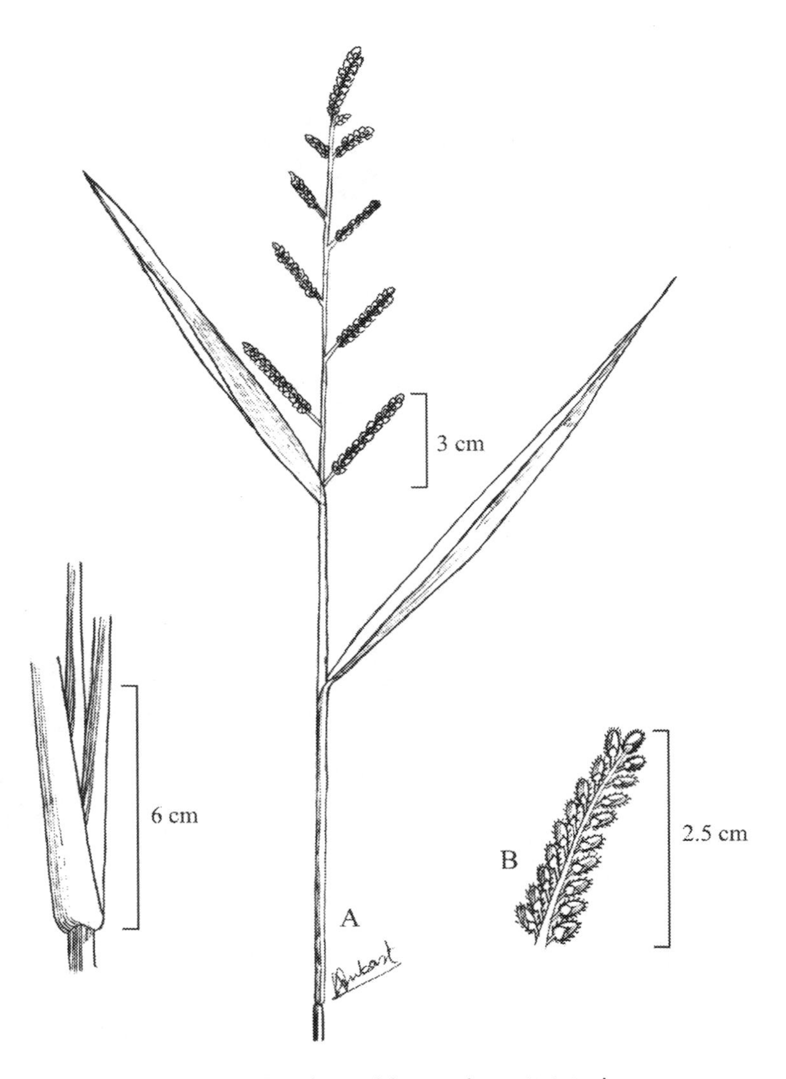

Fig.-18: *Echinochloa colona* (L.) Link

A. Inflorescence

B. Single raceme, the spikelets are more of less hairy

PLATE - XVII

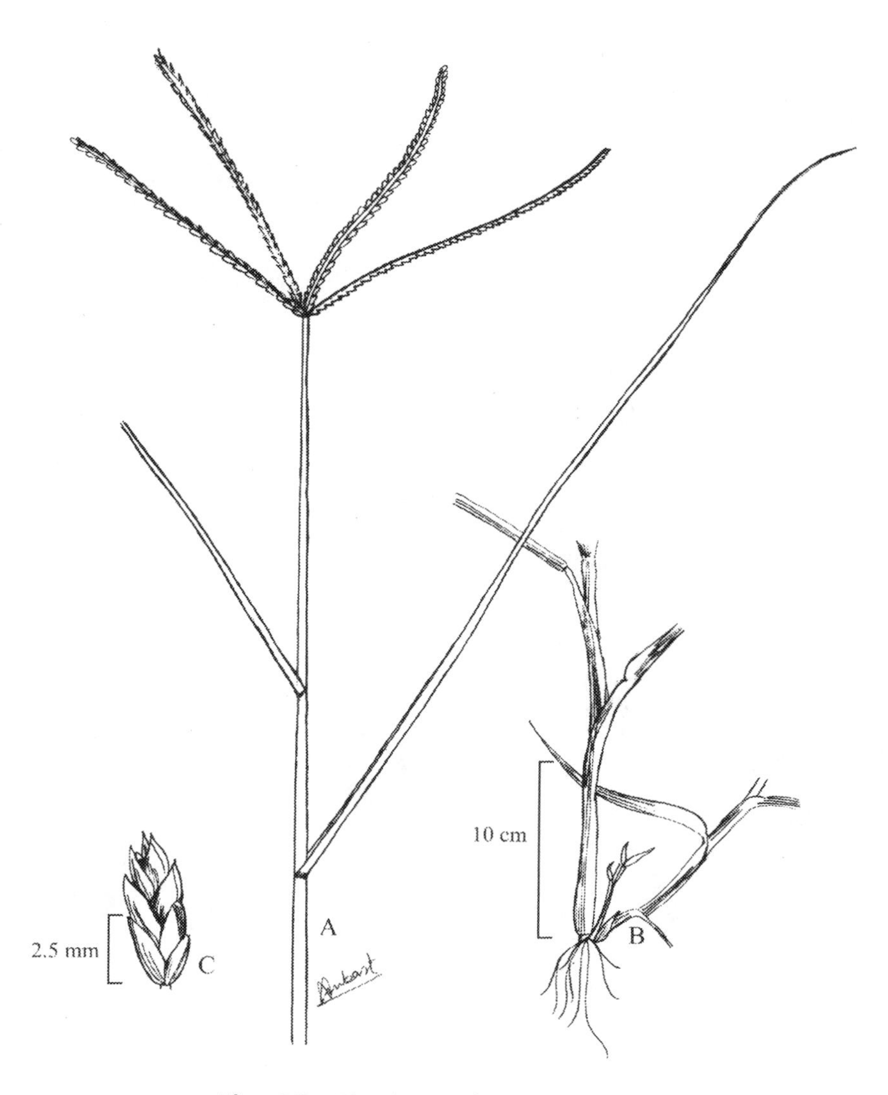

Fig.-19: *Eleusine indica* (L.) Gaertn.

A. Plant
B. Base of stem: an annual
C. A spikelet

PLATE - XVIII

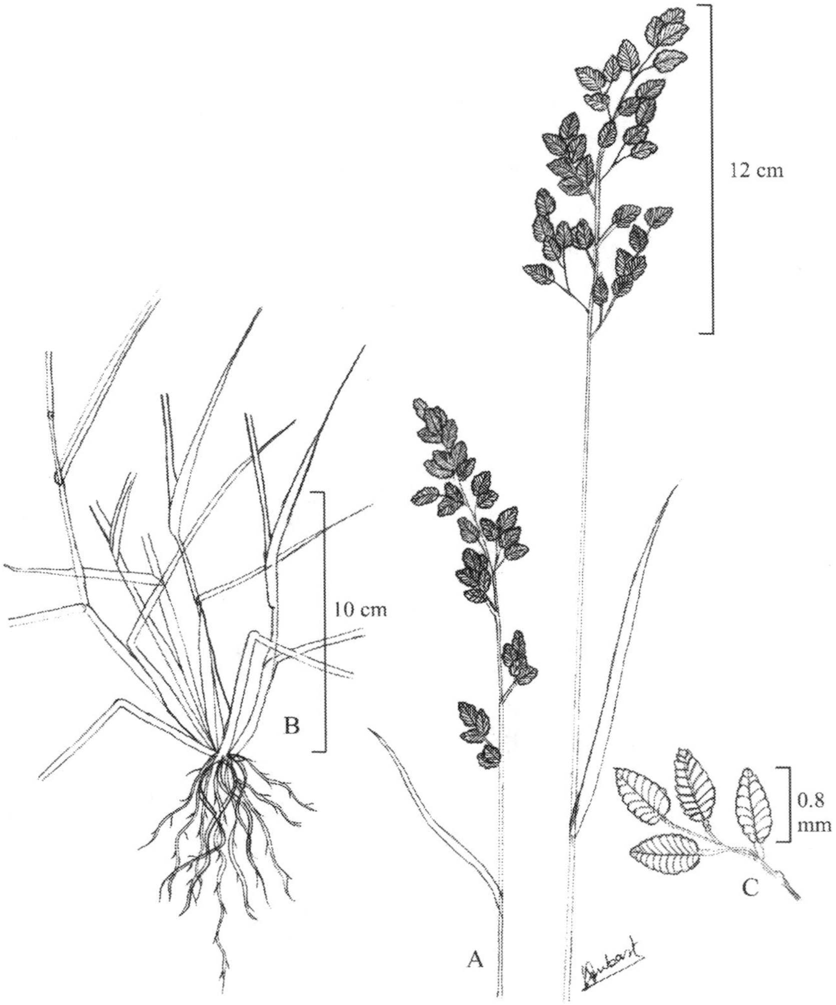

Fig.-20: *Eragrostis unioloides* (Retz.) Nees ex Steud.

A. Inflorescence; the spikelets are reddish
B. Base of plant: annual
C. Spikelets

PLATE - XIX

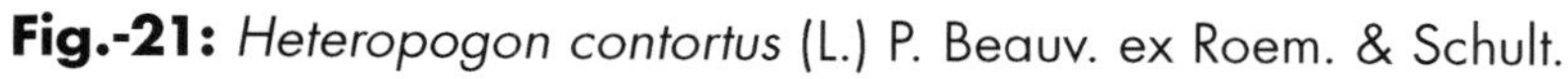

Fig.-21: *Heteropogon contortus* (L.) P. Beauv. ex Roem. & Schult.

A. Upper half of plant showing inflorescence
B. Base of plant
C. Seed with awn attached, showing long twisted hairy awn and sharp pointed callus

PLATE - XX

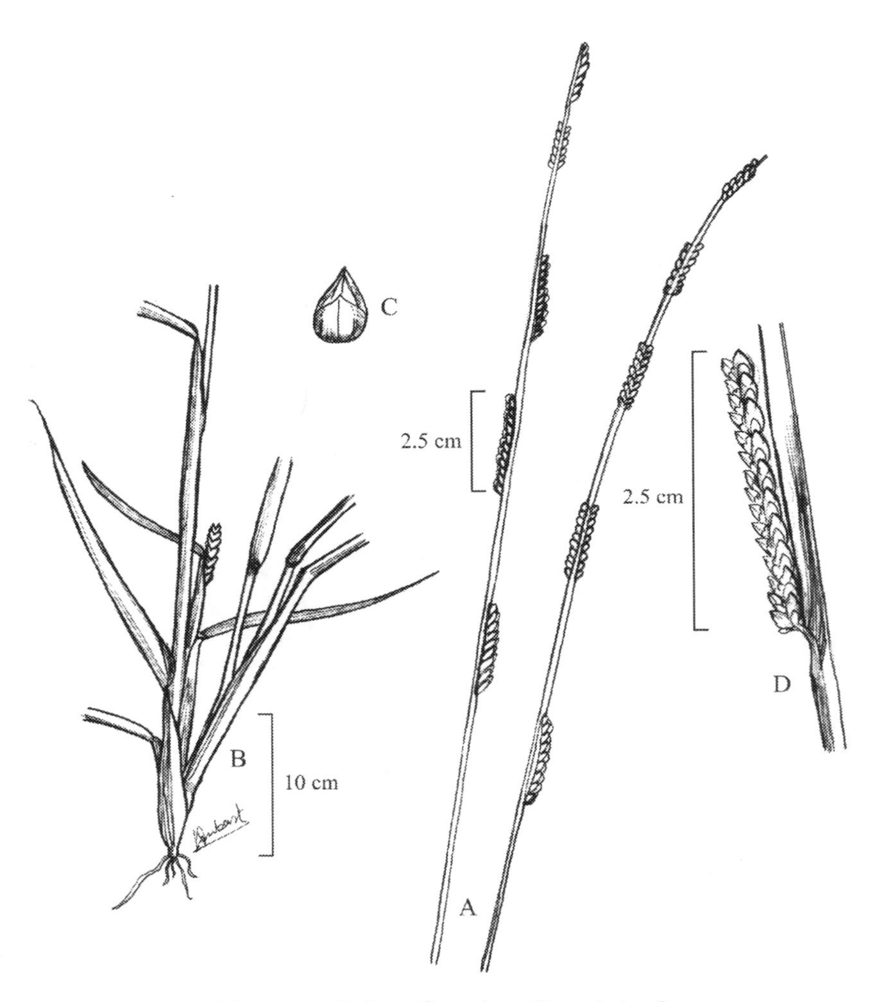

Fig.-22: *Paspalidium flavidum* (Retz.) A. Camus

A. Inflorescence of widely spaced racemes
B. Base of the plants; annual
C. Spikelet, glabrous
D. Single raceme of two rows of second spikelets

PLATE - XXI

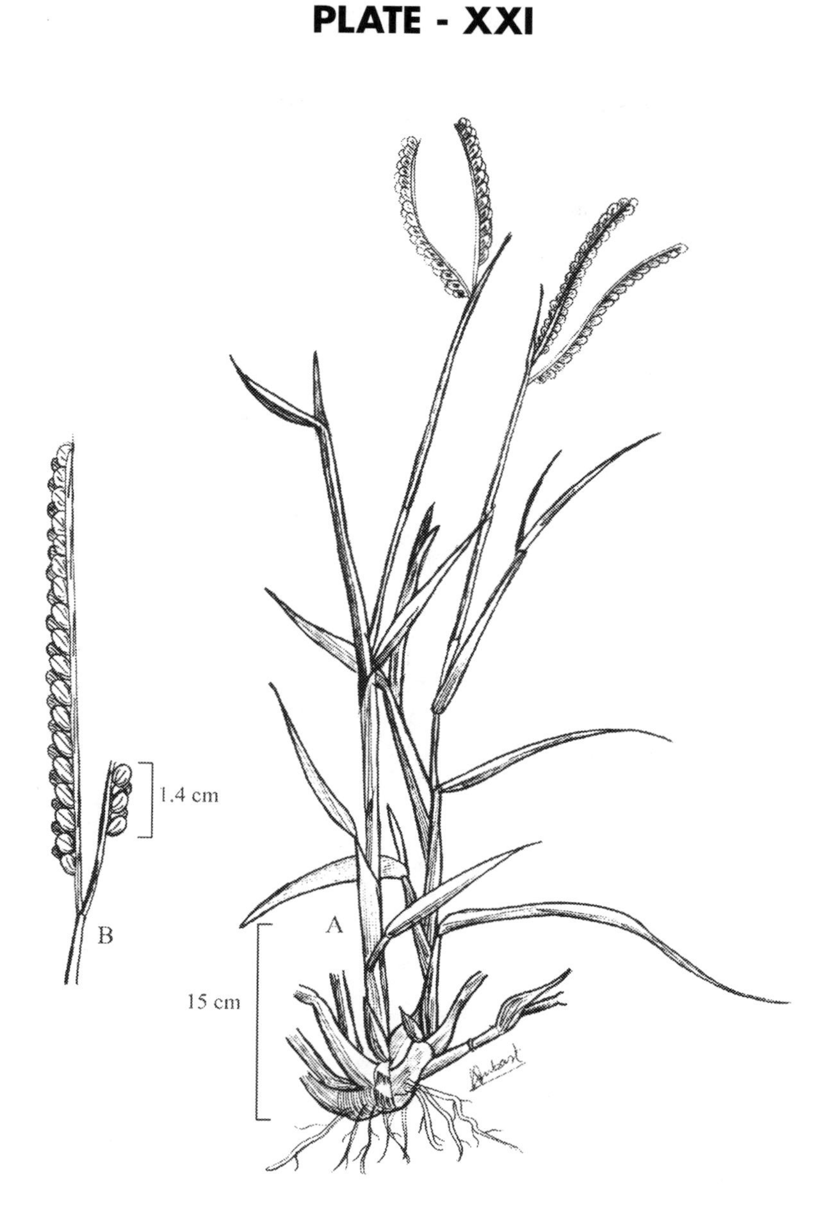

Fig.-23: *Paspalum scrobiculatum* L.

A. Whole plant showing the typical form of the inflorescence of two racemes
B. Raceme, showing the elliptic spikelets in two secund rows

PLATE - XXII

Fig.-24: *Rottboellia cochinchinensis* (Lour.) Clayton

A. Inflorescence, showing the racemes terminal to the culm and its branches
B. A raceme, showing sessile and pedicelled spikelets; the spikelets at the tip of the raceme are imperfect

PLATE - XXIII

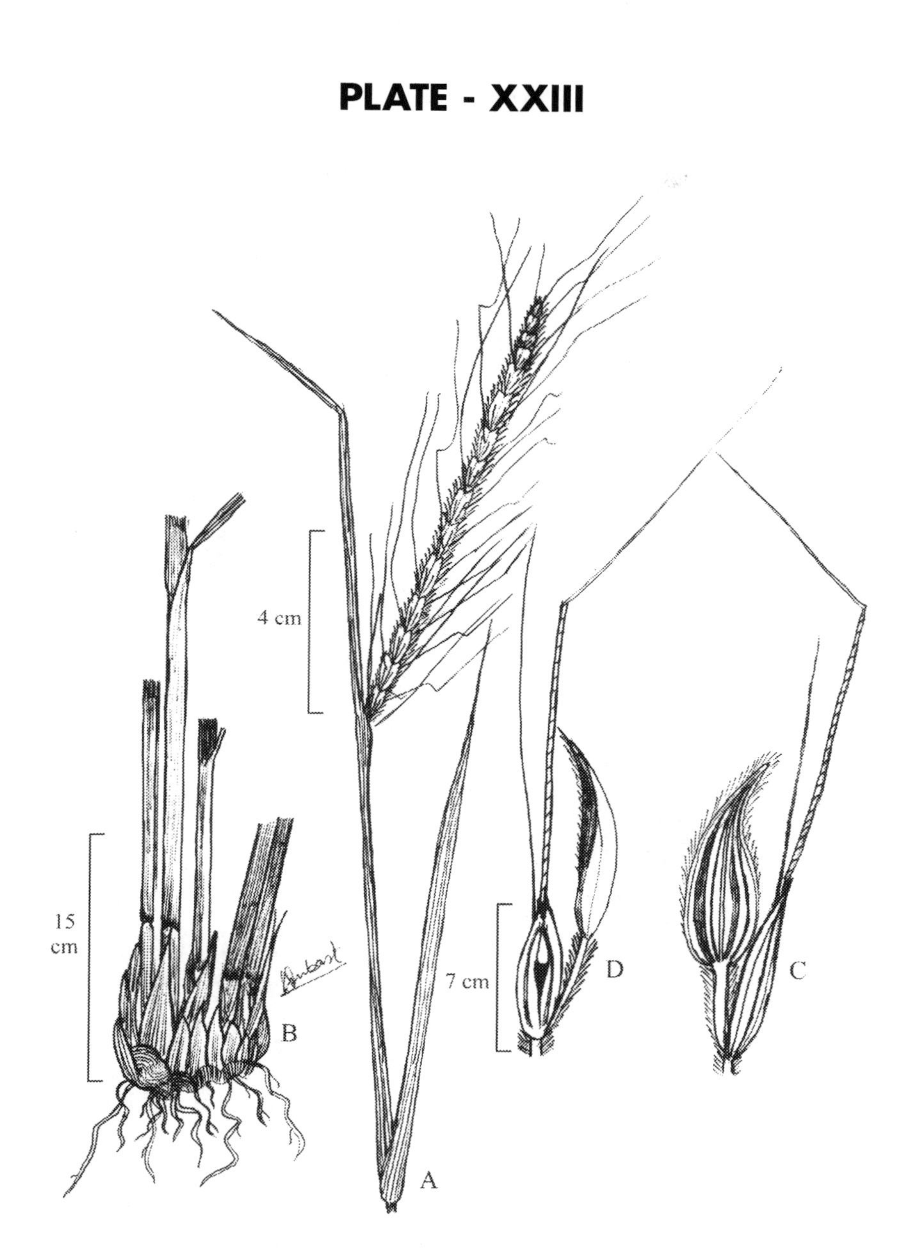

Fig.-25: *Sehima nervosum* (Rottl.) Stapf

A. Inflorescence a solitary terminal raceme
B. Thick rootstock of the plant
C. Sessile awned spikelet in side view, upper glume awned
D. Sessile awned spikelet in full view

PLATE - XXIV

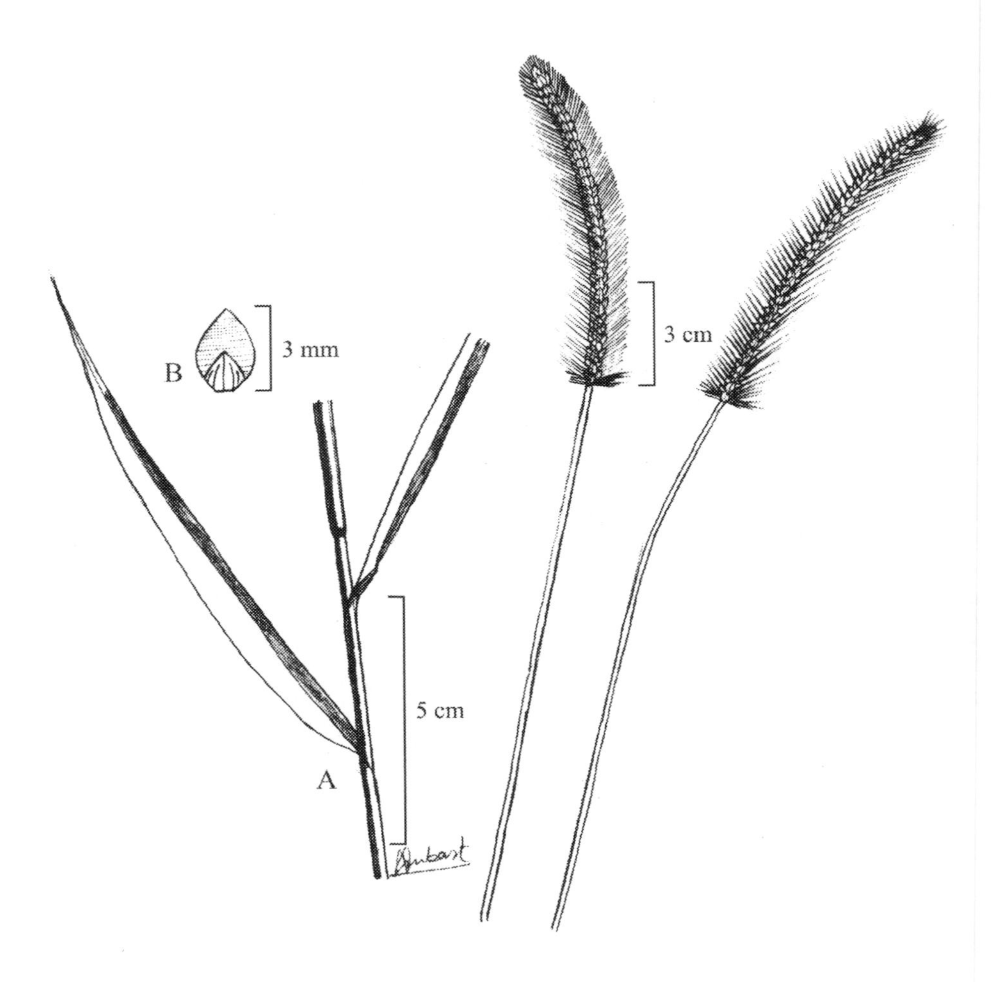

Fig.-26: *Setaria glauca* (L.) P. Beauv.

A. Plant, showing typical terminal racemose inflorescence
B. Back view of spikelet showing the transversely rugose upper lemma

PLATE - XXV

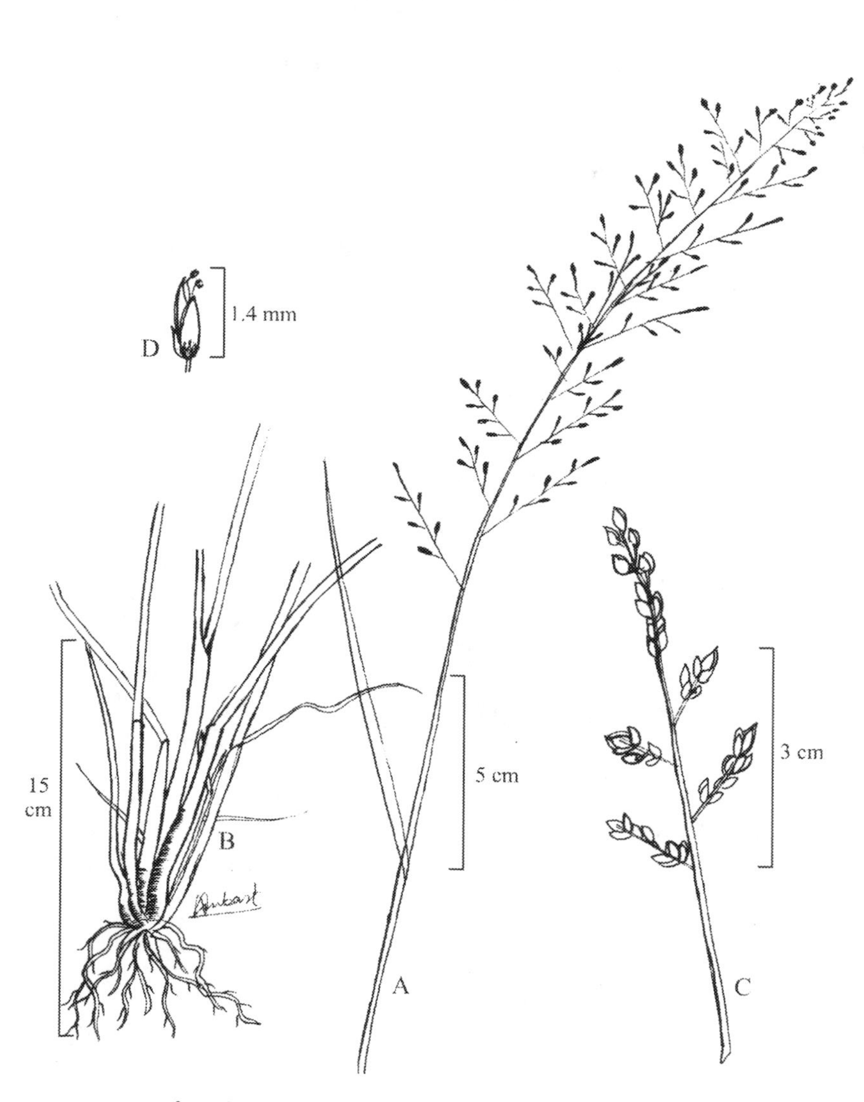

Fig.-27: *Sporobolus diander* (Retz.) P. Beauv.

A. Inflorescence

B. Base of the plant

C. A branch of the inflorescence; note that it is bare at the base

D. A spikelet; glumes both shorter than the lemma and its palea

PLATE - XXVI

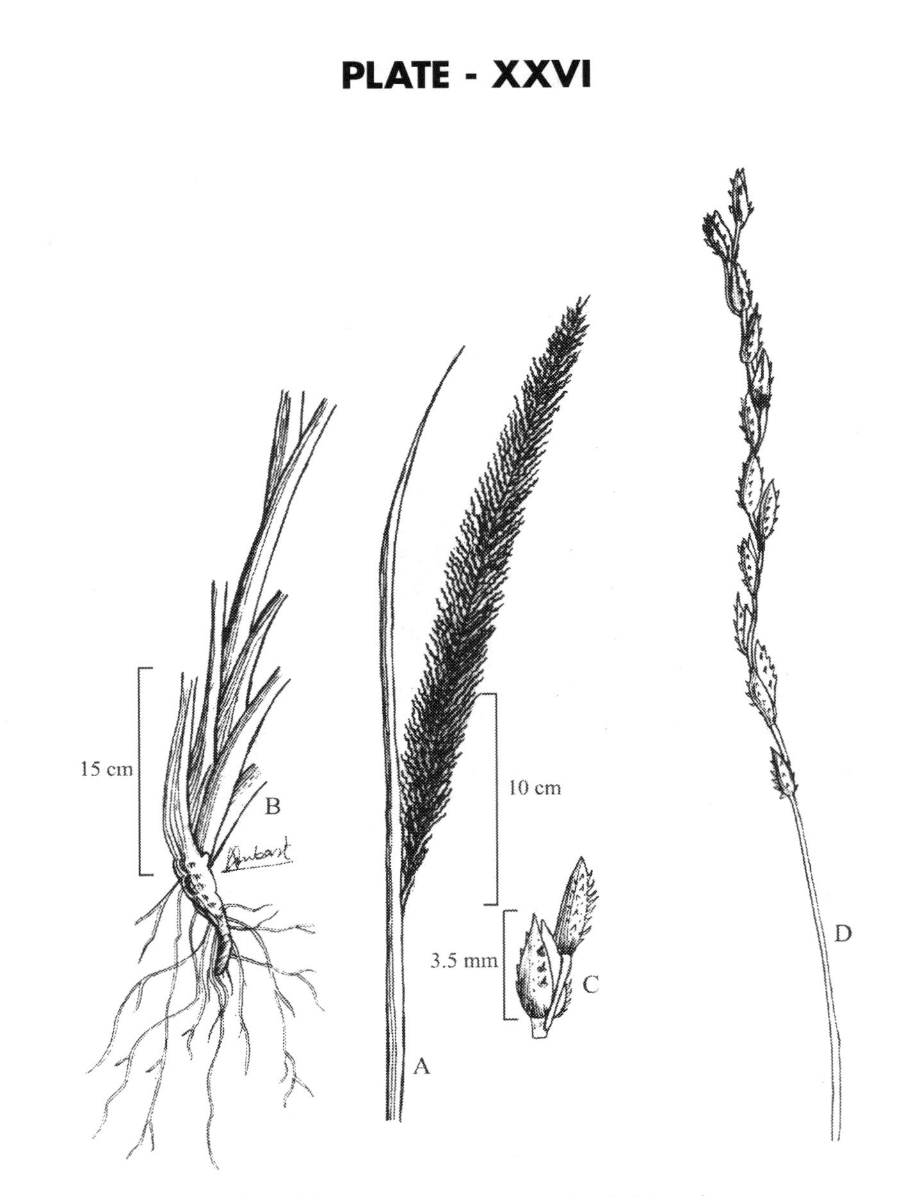

Fig.-28: *Vetiveria zizanioides* (L.) Nash

A. Inflorescence
B. Base of the plants; note the rhizome
C. A short-pedicelled and a long-pediceiled spikelet.
The muricate glumes are characteristic
D. A branch

PLATE - XXVII

Fig.-29 A: Plants of *Cyperus iria* L.
B: Close view of the inflorescence of *Cyperus iria* L.

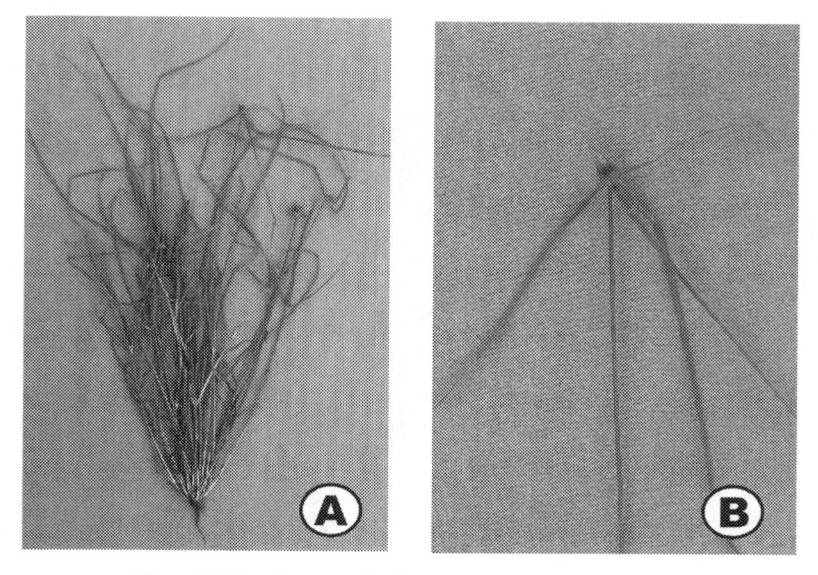

Fig.-30A: Plants of *Cyperus pygmaeus* Rottb.
B: Close view of the inflorescence of *Cyperus pygmaeus* Rottb.

PLATE - XXVIII

Fig.-31A: Patches of *Cyperus rotundus* L. in a Stream of Gautam Buddha Wildlife Sanctuary, Hazaribag

Fig.-31B: Individual Plant of *Cyperus rotundus* L.

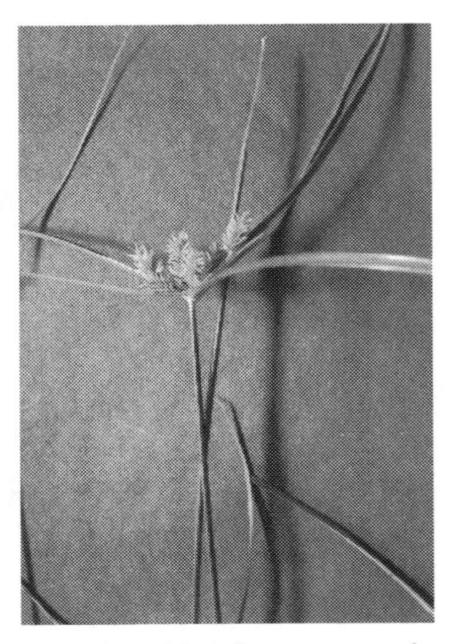

Fig.-32: Inflorescence of *Mariscus paniceus* (Rottb.) Vahl

PLATE - XXIX

Fig.-33: *Aristida setacea* Retz.

Fig.-34: Brooms made from *Aristida setacea* Retz.

PLATE - XXX

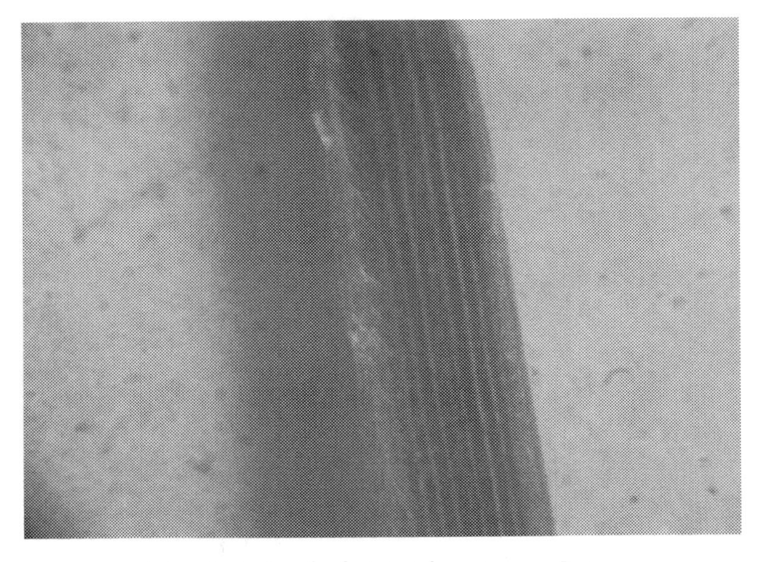

Fig.-35: Leaf of *Aristida setacea* Retz. showing veins and scabrid margin

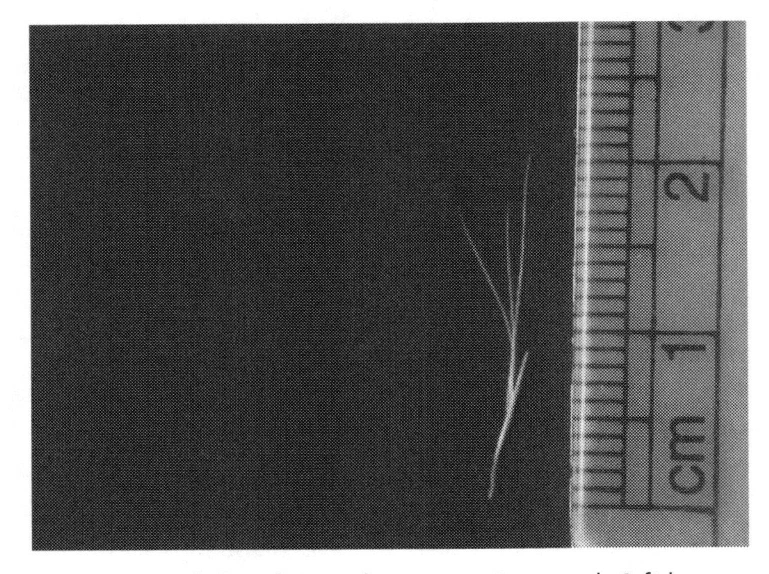

Fig.-36: Spikelet of *Aristida setacea* Retz. with 3-fid awns

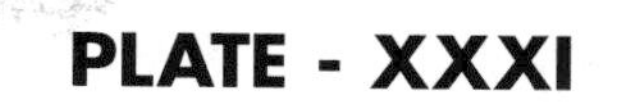

PLATE - XXXI

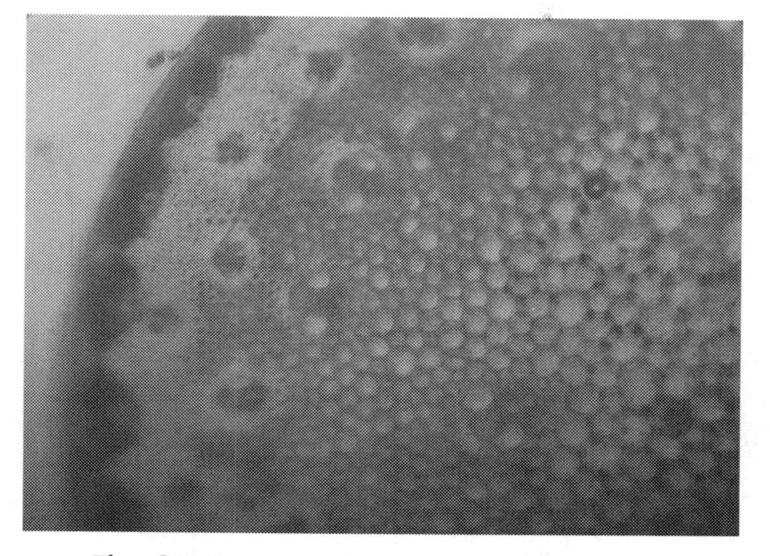

Fig.-37: Transverse Section (T.S.) of the culm of *Aristida setacea* Retz. showing a mass of parenchymatous cells with a number of vascular bundles embedded in it

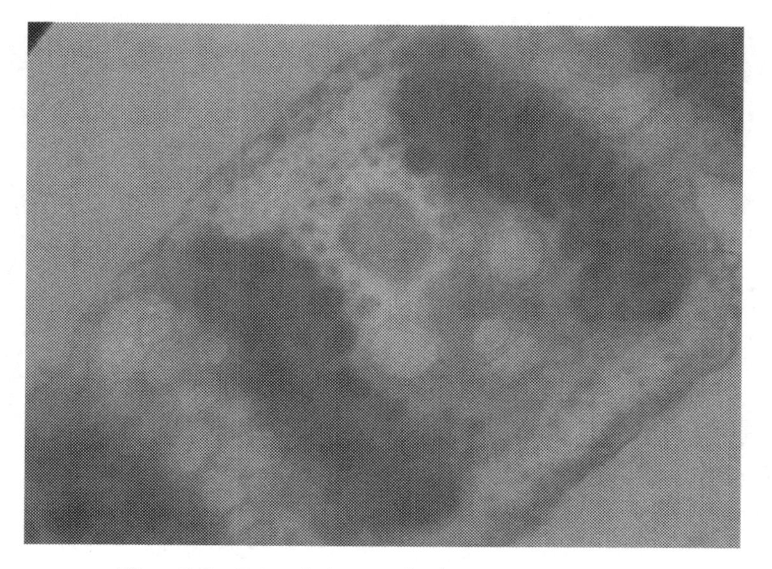

Fig.-38: T.S. of the Leaf of *A. setacea* Retz.

PLATE - XXXII

Fig.-39: *Bambusa arundinacea* (Retz.) Willd.

Fig.-40: Villagers of in and around Gautam Buddha Wildlife Sanctuary, Hazaribag with their products made from *Aristida setacea* Retz. and *Bambusa arundinacea* (Retz.) Willd. at Bigaha Bazar

PLATE - XXXIII

Fig.-41: *Bothriochloa pertusa* (L.) A. Camus

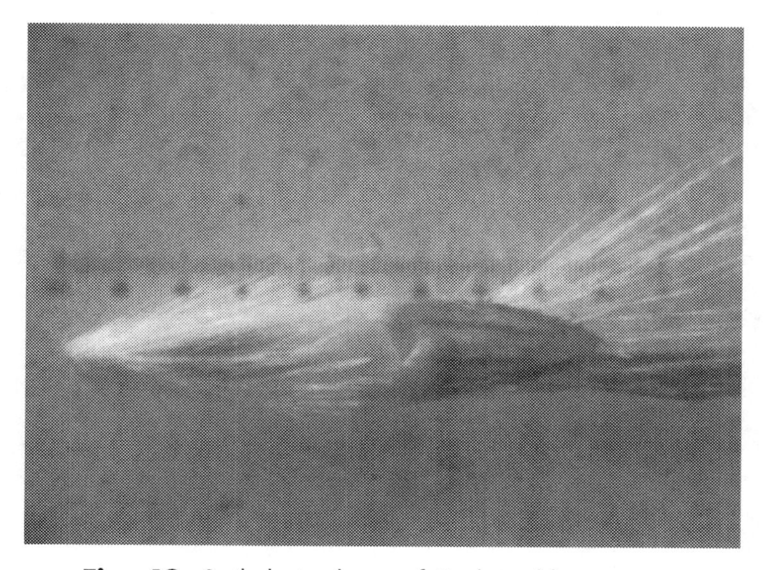

Fig.-42: Spikelet's glume of *Bothriochloa pertusa* (L.) A. Camus with pit and plenty of hairs

PLATE - XXXIV

Fig.-43: *Brachiaria ramosa* (L.) Stapf

Fig.-44: *Chloris barbata* Sw.

Fig.-45: *Chrysopogon aciculatus* (Retz.) Trin.

PLATE - XXXV

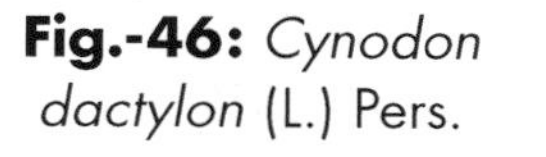

Fig.-46: *Cynodon dactylon* (L.) Pers.

Fig.-47: Inflorescence of *Dactyloctenium aegyptium* (L.) Beauv.

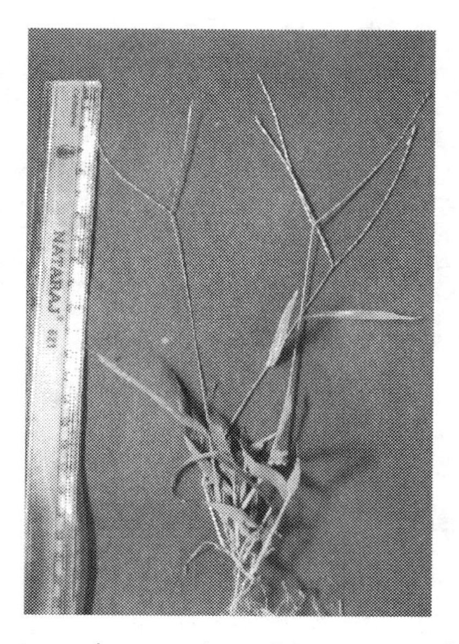

Fig.-48: *Digitaria bicornis* (Lam.) Roem. & Schult. ex Loud.

PLATE - XXXVI

Fig.-49A: *Digitaria ciliaris* (Retz.) Koeler

Fig.-49B: Close view of the inflorescence of *Digitaria ciliaris* (Retz.) Koeler

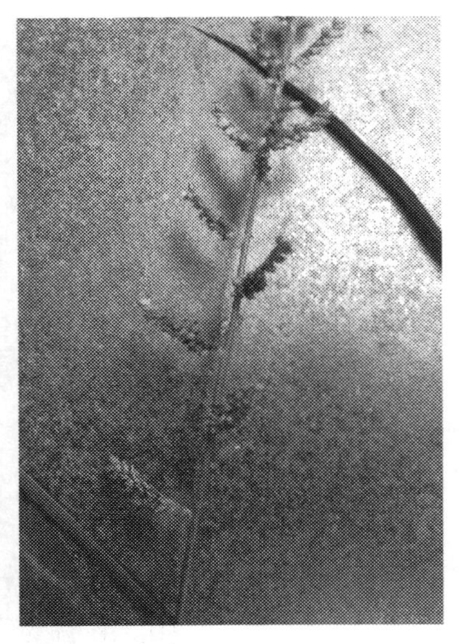

Fig.-50: A part of the inflorescence of *Echinochloa colona* (L.) Link

PLATE - XXXVII

Fig.-51: Showing arrangement of the spikelets in *Echinochloa colona* (L.) Link

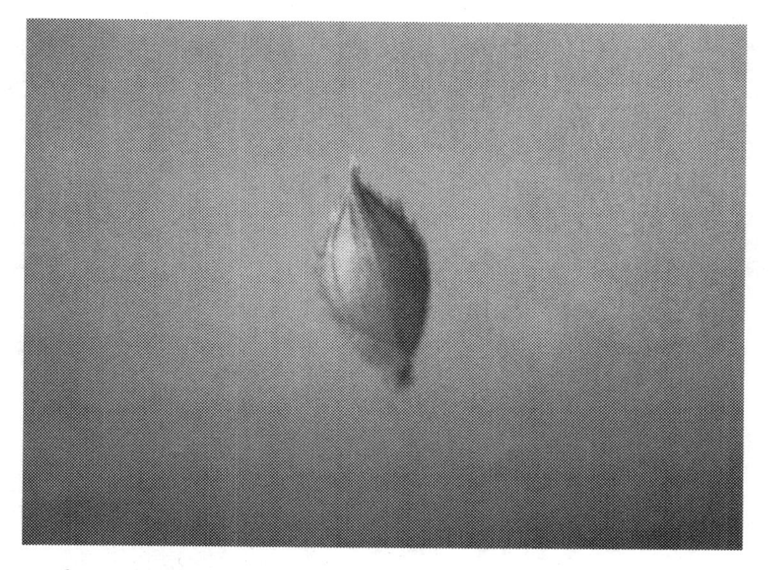

Fig.-52: Single spikelet of *Echinochloa colona* (L.) Link

PLATE - XXXVIII

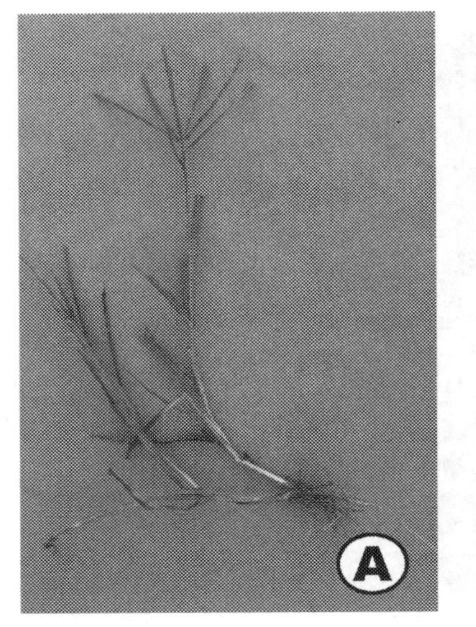

Fig.-53A: *Eleusine indica* (L.) Gaertn.

B: Close view of the inflorescence of *E. indica* (L.) Gaertn.

Fig.-54: Inflorescence of *Eragrostis tenella* (L.) P. Beauv. ex Roem. & Schult.

PLATE - XXXIX

Fig.-55: *Heteropogon contortus* (L.) P. Beauv. ex Roem & Schult., growing on hill

Fig.-56: *Oplismenus burmanii* (Retz.) P. Beauv.

PLATE - XXXX

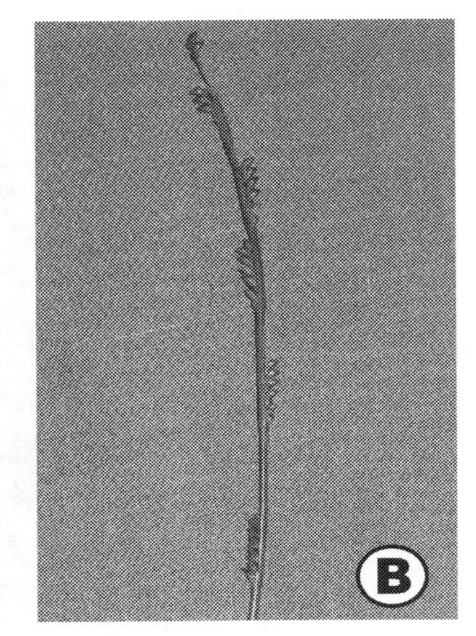

Fig.-57A: *Paspalidium flavidum* (Retz.) A. Camus

B: Close view of the inflorescence of *P. flavidum* (Retz.) A. Camus

Fig.-58: Inflorescence of *Pennisetum pedicellatum* Trin.

PLATE - XXXXI

Fig.-59: Inflorescence of *Phalaris minor* Retz.

Fig.-60: Inflorescence of *Polypogon monspeliensis* (L.) Desf.

PLATE - XXXXII

Fig.-61: *Saccharum bengalense* Retz.

Fig.-62: *Saccharum spontaneum* L.

PLATE - XXXXIII

Fig.-63: *Sacciolepis interrupta* (Willd.) Stapf

Fig.-64A: *Setaria glauca* (L.) P. Beauv.

B: Close view of the inflorescence of *S. glauca* (L.) P. Beauv.

PLATE - XXXXIV

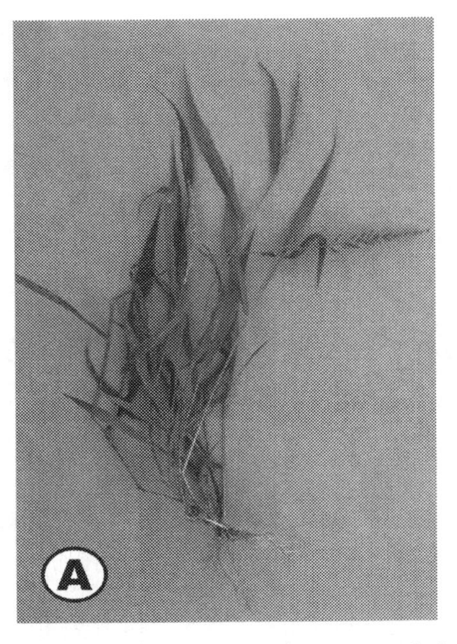

Fig.-65A: *Setaria intermedia* Roem. & Schult.

B: Close view of the inflorescence of *S. intermedia* Roem. & Schult.

PLATE - XXXXV

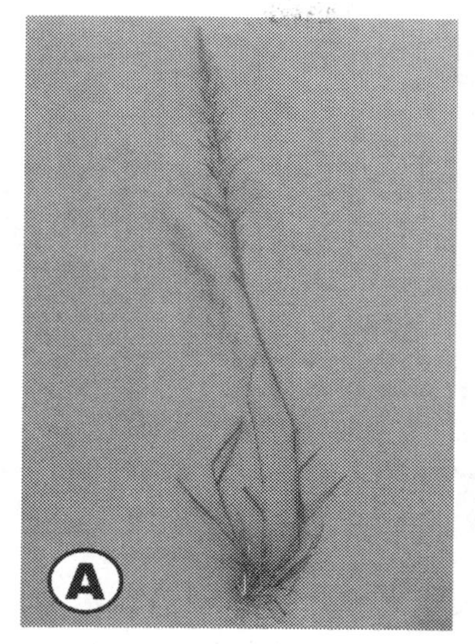

Fig.-66A: *Sporobolus diander* (Retz.) P. Beauv.

B: Inflorescence of *S. diander* (Retz.) P. Beauv.

Fig.-67: *Zea mays* L.

Abstract

Gautam Buddha Wildlife Sanctuary constitutes a part of Wildlife Division, Hazaribag. It is located within Hazaribag District and has a forest area of 100.05 sq. km. It provides a variety of flora and fauna. It's nearest Railway station being Koderma (41 km) and nearest airport Ranchi (160 km).

Due to deforestation and mining activities in and around the area of my study i.e. Gautam Buddha Wildlife Sanctuary, Hazaribag, sedges and grasses account for the largest group of plant in this region. Before we perceive the loss due to the falling of trees, grasses come into the picture and we somehow or the other realize the perception of the flaw in the terrestrial scheme, by which what was good for God's birds was bad for God's garden or that cruelty towards one set of creation (here trees) was benevolence (here grasses) towards another. Within a short span the once thick forest cover are taken over by grasses and sedges and turned into a grazing ground (grassland). However, they have not been adequately studied so far by taxonomists. Therefore, the literature available is scanty.

Cyperaceae or the sedge family with its wide range of distribution and habitat adaptability is adequately represented in this area. On the other hand sedges are mostly perennial or sometimes annual herbs. Because many species have a tufted growth habit, long thin-texture, narrow flat leaves with a sheathing base, a jointed stem and much branched inflorescence of tiny flowers, they are often described as graminoid, meaning grass like. Indeed many horticultural references include sedges under the general heading of grasses. Further more, many sedges and grasses do not fit the graminoid image at all, for example having leaf blades rounded in cross section or no leaf blades at all, or having compact, head like inflorescences. So, we end up with some sedges and grasses being called graminoid, and others being simply called herbs. To minimize confusion we prefer to restrict the term graminoid to true grasses (Family Poaceae) and to apply the new term cyperoid to members of the sedge family.

Grasses are rather interesting in that they are usually successful in occupying large tracts of land to the exclusion of other plants. If we take into consideration the number of individuals of any species of grasses, they will be found to out-number those of any species of any other family. Even as regards the number of species this family ranks fifth, the first four places being occupied respectively by Compositae, Leguminosae, Orchidaceae and Rubiaceae. As grasses form an exceedingly natural family it is very difficult for beginners to readily distinguish them from one another.

In the present study of the sedges of Gautam Buddha Wildlife Sanctuary, it is discerned that the family Cyperaceae is represented by 9 genera as opposed to 12 that are reported in Flora of Hazaribagh District (Paria and Chattopadhyay, 2005); 20 in Flora of Bihar (Singh et al., 2001) and 11 in Botany of Bihar and Orissa (H.H. Haines, 1924). After collecting the data it was observed that the genus *Bulbostylis* has two species namely *B. barbata* and *B. densa.* Paria and Chattopadhyay (2005) reported 1 species. The genus *Carex* has only one species in Gautam Buddha Wildlife Sanctuary namely *Carex cruciata* where as Flora of Hazaribagh District, Flora of Bihar, and Botany of Bihar and Orissa reported 1, 10 and 6 species respectively. *Cyperus* is the largest genus in the sense that it has 15 species in Gautam Buddha Wildlife Sanctuary. The second largest representative genus is *Fimbristylis* with 6 reported from Sanctuary, 10 from Flora of Hazaribagh District, 24 from Flora of Bihar and 23 species from Botany of Bihar and Orissa. The genera *Mariscus, Pycreus* and *Schoenoplectus* have same number of species i.e. 2 in the Sanctuary; in Flora of Hazaribagh District, they are represented by 2, 2 and 3 species respectively. In Flora of Bihar, they have 7, 7 and 11 and in Botany of Bihar and Orissa, they have 6 and 7 in *Mariscus* and *Pycreus* while *Schoenoplectus* was at that time included in *Scirpus* with 8 spp. *Kyllinga* has 3 each in Gautam Buddha Wildlife Sanctuary, Hazaribag and Flora of Hazaribagh District while in Flora of Bihar and Botany of Bihar and Orissa, it has 6 and 4 respectively. It is imperative to note that the genera which have been reported in Flora of Hazaribagh District but not in Gautam Buddha Wildlife Sanctuary, Hazaribag are *Eleocharis, Rikliella* and *Scleria*. The facts gleaned from the data are that altogether 34 species and 2 sub-species have been reported from Gautam Buddha Wildlife Sanctuary. This is not much less than that reported by Paria and Chattopadhyay (2005) in Flora of Hazaribagh

District. Singh et al. (2001) reported altogether 139 species, 3 subspecies and 20 varieties. Haines (1924) reported 100 species and 4 varieties. Altogether 12 sedges which have been reported from Gautam Buddha Wildlife Sanctuary but not from Hazaribag by Paria and Chattopadhyay (2005) are *Bulbostylis densa, Cyperus bulbosus, C. castaneus, C. dubius, C. nutans, C. pangorei, C. pygmaeus, Fimbristylis, alboviridis, F. littoralis, F. polytrichoides, Mariscus compactus* and *M. paniceus*. Out of them *C. bulbosus, F. alboviridis, F. littoralis* and *F. polytrichoides* were not reported by Haines (1924). The species not reported by Singh et al. (2001) are *C. dubius, F. alboviridis* and *F. polytrichoides.*

Altogether 49 genera of Poaceae have been reported from Gautam Buddha Wildlife Sanctuary during this work. The total number of species reported is 76. The most dominant genera are *Eragrostis* with 7 species followed by *Panicum* and *Setaria* each having 4 spp. *Digitaria* and *Saccharum* have 3 spp. each. The other dominant genera are *Sacciolepis, Aristida, Bothriochloa, Brachiaria* and *Dichanthium*. In confirmity to my findings Paria and Chattopadhyay (2005) also marked *Eragrostis* (10 spp.) as the largest genus of Hazaribag District. *Panicum* (5 spp.), *Setaria* (4 spp.) and *Digitaria* (4 spp.) were also the other three major genera which again confirmed my findings. However *Cynodon* (3 spp.) which is the 5th largest genus from Hazaribag has only one species *C. dactylon* in Gautam Buddha Wildlife Sanctuary. Altogether 13 (thirteen) rare grasses which have been reported from Gautam Buddha Wildlife Sanctuary and not by Paria and Chattopadhyay (2005) are *Aristida adscensionis, Bothriochloa pertusa, Brachiaria reptans, Chloris dolichostachya, Desmostachya bipinnata, Digitaria bicornis, Iseilema anthephoroides, Paspalidium geminatum, Sacciolepis interrupta, S. myosuroides, Sehima nervosum Setaria pumila* and *Vetiveria zizanioides.*

In course of study grasses of great ecological significance have been reported. *Cynodon dactylon, Desmostachya bipinnata, Eleusine indica, Aristida adscensionis* and *Echinochloa colona* were found to be drought resistant. Some of the grasses were good soil binders. They are *C. dactylon, D. bipinnata, E. indica, Panicum paludosum, P. psilopodium* etc. The grasses which fought soil erosion on hilly tracks are *C. dactylon, Cymbopogon martinii, Dactyloctenium aegyptium* etc. Most significant finding was identifying grasses which colonized left over mines dunes. Some of them being *A. setacea, Saccharum spantaneum, S. bengalense, A. cimicina, Digitaria ciliaris* and *D. longiflora*.

The identification of the grasses on the basis of their vegetative parts is both unscientific and erroneous. It also goes against the inviolable rule that the basis of classification of the grasses is spikelet. However with certain degrees of dexterity and intelligence one may be trained to identify certain grasses which we frequently come along in my area of study. It is in this light that we prepared an Artificial Key (Bor, 1941) and assigned grasses found in Gautam Buddha Wildlife Sanctuary to 10 groups (A – J). The Key has been kept as simple as possible and where it seems possible to make a mistake in relegating a species to one group or another, that species is included in both groups. All the same the key does not obviate the study of microscopic characters altogether and that the use of hand lens and microscopic study together with the study of macroscopic characters have been included in the preparation of the key.

The involvement of the governmental agencies in tackling ecological perils has long been overemphasized but total dependency on them is again foolhardy. Also the degree of dependence on NGOs or Local bodies in these efforts is a vexed problem. This has led to a situation where neither is ready to share responsibilities. This in turn has left our precious green cover to the mercy of contractors. Tired of official obfuscation on the government side, the self help groups if they ever exist have in recent years raised up their ante against the colosal loss of the green Pagoda. The self-help operation has a positive role to play. Criticizing the increasing apathy of the establishment is not going to alleviate the situation; rather it would be detrimental in obtaining their much needed favour and help. The grass lands are definitely not the panacea for the lost canopies but certainly they have done their bit to save the nature from cataclysm and catastrophe.

We have at least four reasons for wanting to keep grasses out of harm's ways:

- To avoid the denuded forest turning into barren lands. Somehow or the other grasses tend to survive and flourish in denuded tracts.
- Grasses sustain the animal wealth of the area. Both domestic and wild grazers are dependent upon them.
- Many grasses are excellent soil binders and they check the precious soil from being swept off due to rains, flood and air storms.
- Some of the grasses are proven hearth burners and bread earners. They are being used fairly commonly in house hold activities, and in cottage industries.

Thus it is a racing certainty that unless and until something is quickly done we will continue to witness the perils of unplanned growth. Taxonomists here have a role to play and study the neglected plants such as sedges and grasses are all the more important to replenish our knowledge base about them. Our floral wealth is being cornered with pressure mounting on it from within the society and from without. It would indeed be an egregious mistake to take things lying down.

Introduction

The members of the family Cyperaceae and Poaceae with their wide range of forms, types and habitat enjoy a sway over most areas of growth and covers. They comprise important undergrowth of forests and land of significance in various ways. They are the pioneers of deforested and denuded tracts of biosphere. The perception in the flaw in the terrestrial scheme of things by which what is bad for one set of creature is good for another set. The ecological loss in terms of forest cover (trees and shrubs) becomes a boom for sedges and grasses. They are also the least affected by natural vagaries. Their fragile and uninhibited frames are to their benefit rather to their blame.

Due to deforestation and mining activities in and around the area of the study i.e. Gautam Buddha Wildlife Sanctuary, Hazaribag; Sedges and grasses account for the largest group of plants in this region. However, they have not been adequately studied so far by taxonomists. Therefore, the literature available is scanty. A comprehensive work for this region is still awaited. Recently, taxonomists like Paria and Chattopadhyay (2005) have thrown some light on them but still much is awaited. Our endeavor here is to study them in detail.

Cyperaceae or the sedge family with its wide range of distribution and habitat adaptability found a place even in the pre-Linnaean contributions of Tournefort (1719) and Micheli (1729). Linnaeus (1753 and 1754) described 5 genera and 81 species. Subsequent floristic works included many novelties in Cyperaceae also, and of these historical works, in particular relation to Indian plants, mention may be made of Burman (1768), Linnaeus (1767 and 1771), Rottboell (1773), Retzius (1786-1791), Willdenow (1797-1830), Vahl (1805/1806), R. Brown (1810), Roxburgh (1820, 1824, 1832) and Miquel (1855-1859). These were followed by more comprehensive treatments of Cyperaceae by Nees (1834), Kunth (1837), Steudel (1854-1855), Boott (1858-1867), Boeckeler (1868-1877), Bentham (1881-1883) and Pax (1888). However, it is Clarke's pioneer contribution (1893, 1894) in J.D. Hooker's Flora of British India which even after 90 years still continues to be the main source of information on Indian Cyperaceae. The accounts of the family in Indian Regional Floras by Prain (1903), Cooke (1909), Haines (1924), Parker and Turill (1929) and Fischer (1931) are largely based on Clarke's contribution.

Some of the specimens deposited in the Central National Herbarium, Botanical Survey of India, Howrah were also consulted. In all, the treatment includes genera and species of Cyperaceae of India and the world.

A large cosmopolitan family of most herbaceous plants, Cyperaceae occurs primarily in most temperate to wet tropical regions of the world, several species are of economic importance.

Sedges are mostly perennial or sometimes annual herbs. Because many species have a tufted growth habit, long thin-textured, narrow flat leaves with a sheathing base, a jointed stems and much-branched inflorescences of tiny flowers, they are often described as graminoid, meaning grass-like. Indeed many horticultural references include sedges under the general heading of grasses. Furthermore, many sedges and grasses do not fit the graminoid image at all, for example, having leaf blades rounded in cross section or no leaf blades at all, or having compact, head-like inflorescences. So, we end up with some grasses and sedges being called graminoids, and others being simply called herbs. To minimize confusion we prefer to restrict the term graminoid to true grasses (Family Poaceae), and to apply the new term cyperoid to members of the sedge family.

Along with the similarities to grasses there are many features that distinguish sedges (such as usually leaves arranged in threes, and usually a solid stem that is triangular in cross section and usually a conspicuously bracteate inflorescence), but they are not all observable by necked eye. As in grasses, the basic unit of the inflorescence in sedges is the spikelet. Within the family there is enormous variation in spikelet and inflorescence structure. The spikelet consists of one to several, tiny, male, female or bisexual flowers, each born in the axil of a boat-shaped glume (tiny bract) which is coloured in shades of green or brown, red, occasionally white or bright yellow.

The largest genus is *Carex* with about 2000 species world-wide, followed by *Cyperus* with about 550 species. Sedges occur primarily in the tropics and sub tropics, but may be locally dominant in some areas like the sub arctic regions (Tundra).

Sedges are found mainly in wetlands (some are entirely aquatic) and along water courses, but also occur in moist grasslands and along forest margins.

Grasses are evenly distributed in all parts of the world, acquire wide tracts of land. They are found in every kind of soil and in all kinds of situations and climatic conditions. Somewhere grasses form a leading feature of the flora. Grasses generally grow in open places and so they are not usually abundant under the shades of the trees either as regards the number of individuals, or of species. But in open places they become abundant and sometimes convert whole tracts into grass-lands. So, most of the area of the actual vegetation would consist of grasses..

Grasses are of great importance to human being on the basis of their almost universal distribution and their great economic value. Still very less number of people appreciates worth of the grasses. Several families of plants supply the wants of human being but Poaceae (Gramineae) exceeds all the others in the amount and the value of its products. The grasses which grow in pasture land and the cereals grown all over the world are of much more importance to human and his domestic animals than all other plants.

Generally we think that grasses include only herbaceous plants with narrow leaves but it also includes the cereals, sugarcane and bamboos.

The interesting feature of the grasses is that they can easily occupy large tracts of land to the exclusion of other plants. The number of individuals of any species of grass will be found to out-number those of any species of any other family. In regard to the number of species this family ranks fifth, the first four places being occupied by Compositae, Leguminosae, Orchidaceae and Rubiaceae respectively.

It is very difficult for beginners to readily distinguish grasses from one another. The leaves and branches of grasses are very much similar; flowers are so small in the way that they are liable to be passed by unnoticed. Even the recognition of our common grasses is a very difficult task for a botanist.

The Gramineae (Poaceae) is the most widely distributed and is in greater abundance than any other group of flowering plants. Species belonging to this family are usually known as grasses. Grasses are distributed in various climates, soils and elevations on the surface of the globe. They have been recorded in marshes, deserts, prairies and woodland, on sand, rock and fertile, alkaline and saline soils from tropics to the polar regions and from sea level to altitudes of perpetual snow.

The Gramineae is one of the largest families of flowering plants. The number of genera is a matter of diverse opinion. It may be 620 genera (Hutchinson, 1959) and about 10,000 species in the world (Hubbard, 1954). About 240 genera are represented in India (Bor, 1960). About 1,200 species are supposed to be occurring in India (Jain, Doli Das and Banerjee, 1972). For western India, Blatter and McCann (1934) reported 284 species. Fischer C.E.C. (1934 and 1936) published 377 species from South India. Haines (1924) recorded 201 species in East India. For Marathwada area of Maharashtra 180 species were recorded (Patunkar, 1980). Raizada, Jain and Bharadwaja (1957, 1966) have reported 286 species for the Upper Gangetic Plain. Bor (1940) included 156 genera and 424 grass species for Assam State. Bor (1941) also described 92 of the commonest U.P. grasses.

J.D. Hooker (1896) reported about 135 genera, 734 species, 2 sub-species and 39 varieties of grasses from the areas of the present day India. Bor (1960) reported about 220 genera and 1165 infrageneric taxa (32.7%) endemic to this continent. Karthikeyan et al. (1989) stated that 16 genera, 299 species, 2 sub-species and 43 varieties are found to be endemic.

According to Nair (1991) there are about 141 endemic genera distributed over 47 families in India and Acanthaceae and Gramineae have the largest number of endemic genera. It is found that about 16 genera of Gramineae are endemic to India (Shukla, 1996).

The fourth largest family of flowering plants i.e. Gramineae in the world with over 700 genera and probably 10,000 species (Sreekumar and Nair, 1991). In the present work 76 species have been recorded. Most of the specimens were identified and checked by the author. Flora of British India (Hooker, 1896-1897) and Grasses of Burma, Ceylon, India and Pakistan (Bor, 1960) were consulted for the identification of each species. Therefore, these references are not repeated unless required in case of some species.

The evolution of man and his present position in the biological world has been significantly affected by grasses. The historical records suggest that most of the world's civilizations developed around the regions of grassland. It would not be unfair to say that the human population has attained its present level of civilization and development due to abundance and widespread distribution of grasses on this earth. Further, it may be added that without grasses not only human population but even the very survival of animals also seems to be obscured.

Much earlier than the recorded history, grasses have provided food, shelter, medicine and sports for man. Domestic animals and many types of Wildlife are directly or indirectly dependant upon grasses and grassland for food, shelter and even for the completion of their life cycle. There are many animals who cannot exist without extensive grasslands such as deer, wild dogs, fox, jackals, rabbits, wild cow (Antelope) and also birds like peacock, wild hens, ducks, pigeon, pheasants, etc. It has been observed in the forests that it is hard to find any animal or bird in teak or sal forest areas. As the forests are being cut and grassland taken under cultivation, most of the wild animals and birds are moving towards the mixed forest areas or becoming extinct. Animals like lion, tiger, leopards and bear are also indirectly dependant on grasslands for food.

The first paper dealing with grasses was published in 1708 by Johann Scheuchzer, entitled Rostographiae Heevetica Prodromus. This may be considered as the beginning of Agrostology, the science of grass classification. Linnaeus (1753) listed only a few genera like *Panicum, Hordeum, Triticum* and *Phalaris.* During the 19th century, there was a general shift in the objectives of systematics to the grouping together of morphologically similar plants; this came to be known as natural classification.

Robert Brown (1810) was the first to understand the true nature of the spikelet and to recognise it as a reduced inflorescence branch. Linnaeus had interpreted the spikelet as a single flower. Brown clearly recognised the two sub-families of Gramineae, the Panicoideae and Pooideae. He described the spikelet characters of these groups very accurately and also noted the tropical and sub-tropical distribution of the former as contrasted with the temperate adaptation of the latter.

In 1812, Palisot de Beauvois described and named a large number of genera. Kunth (1833) distinguished 13 tribes without recognition of sub-families. Among the natural systems Bentham (1881) received much recognition. In his system, 13 tribes grouped into two sub-families namely Panicoideae and Pooideae were recognised, mainly based on morphological characters of the inflorescence and flower. This treatment was presented in Genera Plantarum by Bentham and Hooker (1883) and was followed by Hackel (1889), Stapf (1934), Hitchcock (1920; 1933) and Bews (1929) with some modifications.

In 1931 the publication "Cytotaxonomic investigation in the family Gramineae" by N.P. Avdulov marked the beginning of a new era in grass systematics. In his paper 232 chromosome counts of grasses were correlated with leaf anatomy, first seedling leaf, organisation of the resting nucleus, nature of starch grains in fruit, and geographical distribution of species. Further, he subdivided Poacae into Festuciformes and Phragmitiformes. In 1932, Prat emphasised the significance of leaf epidermis in grass classification and in 1936 he recognised three sub-families; Festucoideae, Panicoideae, and Bambusoideae. Characters correlated in grouping tribes and genera were leaf epidermis and anatomy, cytology, morphology of seedlings, embryos, fruits, inflorescence, nature of starch grains, physiology, ecology and serology. Systems by Prat and Avdulov are phylogenetic where grasses are arranged according to genetic and evolutionary relationships.

In 1954, C.E. Hubbard published his work which supported Avdulov and Prat's views. Between 1950 and 1960 much has been accomplished regarding grass systematics, and attempts were made to collect and correlate the results leading towards the phylogenetic arrangement of the major groups of Gramineae. Six groups of grasses based on characters of leaf anatomy. Notwithstanding all these works, a total arrangement of genera in tribes and tribes in subfamilies has not yet been achieved, however much progress has been done. Most of the agrostologists accept Panicoideae, Festucoideae, Bambusoideae, Eragrostoideae (Chlorideae) as separate sub-families. Prat, Stebbins, Crapton and Parodi have recognised two more sub-families, i.e. Oryzoideae (Pharoideae) and Arundinoideae (Phragmitifomes, Phrag-mitoideae).

Phylogenetic consideration of Gramineae

In general for angiosperms, and particularly for grasses the fossil records are too incomplete and inadequate to help in drawing conclusions on phylogenetic relationships. It has been predicted that grasses came into being during the mesozoic period after the flowering plants were well established and diversified. Further circumstantial evidences support that grasses were evolved in tropics. Recently, Arman Takhtajan (1969) from Russia and Arthur Cronquist (1968) from U.S.A. have outlined their phylogenetic classification of flowering plants. Both agree that the morphological and taxonomic evidences lead to the conclusion that flowering plants had monophyletic origin. It is generally believed that angiosperms were probably derived from seed ferns and monocots from primitive dicots (Ranalian complex). However, there is no complete system dealing with all families and has gained general acceptance. Nevertheless, it is clear that within the monocots, grass flowers have got reduced during the evolutionary course. It is not a simple structure as considered by some workers earlier. Therefore, the Gramineae is a very advanced family rather than being primitive. The order Graminales includes the single family Gramineae.

Morphology of Grass Plant

In general grasses look simple with vertical or horizontal cylindrical stem bearing nodes and internodes. Usually internodes are hollow but in many species of Andropogoneae they are solid. Leaf blades are linear, flat, their basal portion encircles the stem that is known as sheath. Phyllotaxy is two-ranked, i.e. odd numbered leaves are above one another in straight line and even numbered are opposite to each other. Inflorescence is made of one to many spikelets. These spikelets consist of glumes which enclose flowers in the axils.

The grass family has tremendous diversity of shape, size, texture, adaptation and modification of vegetative and reproductive parts. The simple floral structures of Gramineae have been modified during the course of evolution to produce numerous forms. Keen observation will reveal the fact that there are every possible permutation and combination of characters in this family and it is a very complicated and difficult family for taxonomists to deal with.

Duration

Annual or perennial habits of grasses have been considered as important diagnostic characters though sometimes it is difficult to determine the habit from dried herbarium specimens. Some species may be annual in dry conditions and behave as perennial under moist conditions. Same is true for tropics and temperate climates. Regardless of annual or perennial habit, the flowering culms in most of the grass species die off after fruiting. However in some species the culms persist for several years as in *Imperata* etc.

General features of Poaceae

Annual to perennial herbs, woody shrubs or trees like bamboos; roots fibrous; rhizomes present or absent, sometimes stolons developed. Clums erect, ascending or prostrate and creeping; internodes usually hollow but solid in Andropogoneae and some other groups too; nodes, usually terete, sometimes compressed or angled. Leaves usually solitary at the nodes, may be crowded at the base of culms, alternate and two-rowed consisting of sheath, ligule and blade; sheaths encircle the culms, with free or over-lapping margins, frequently swollen at the base; ligule placed at the junction of the sheath and blade, membranous, chartaceous or reduced to a fringe of hairs, rarely absent; blades usually flat, linear to lanceolate, venation parallel, rarely with a constricted petiole.

Inflorescence a panicle, raceme or spike. Basic unit of grass inflorescence is a spikelet, sessile or pedicelled, with one to more sessile florets, its axis is rachilla, continuous or jointed. At the base of spikelet are two empty glumes, lower and upper. Above these glumes are florets. Each floret consists of lemma and palea called the floral bracts. In the axil of the palea the flower is enclosed, bisexual or unisexual. The highly modified and reduced perianth is represented usually by 2 lodicules, stamens are usually 3, sometimes 2 or 6. Pistil with superior ovary, unilocular, one-ovuled, style 2, stigmas plumose. Fruit 1-seeded, indehiscent, rarely with mucilage forming perianth.

INTRODUCTION TO THE AREA

Hazaribag District of Jharkhand has an area of 4,578 sq. km. (Plate-I; Fig.-1 and 2). The District has two sub-divisions Hazaribag and Barhi and 11 blocks namely. Keredari, Barkagaon, Katkamsandi, Churchu, Hazaribag, Barhi, Ichak, Vishnugarh, Barkattha, Chouparan and Padma. The district was established in the year 1833. Literarily Hazaribag mean "Land of thousand gardens" also the name might have been derived from a place "Hazari" located close by. The first municipality was constituted in 1886 and till 1919 it was headed by Englishmen. The District was divided on 6th December, 1972 and Giridih was separated from it. Later from time to time 3 new Districts were carved out of it namely Chatra, Koderma and Ramgarh. As per 2001 census the population of Hazaribag is 14,37,993.

1. Name, Location, Constitution and the Extent of Area

The research area named as Gautam Buddha Wildlife Sanctuary, Hazaribag is situated in Hazaribag District of Jharkhand (Plate-II; Fig.-3). Gautam Buddha Wildlife Sanctuary has derived its name from Lord Gautam Buddha. Areas in and around the sanctuary has been associated with Buddha. Many remnants of ancient Buddha relics lie spread throughout the area. The boundaries of Bodh-Gaya, the famous shrine associated with Lord Buddha and Gautam Buddha Wildlife Sanctuary are more or less contiguous. The Bhadrakali temple at Itkhori is a historical site of significance for Hindu, Budhist and Jain. The Sanctuary is elongated in shape and can be divided into two halves, Northern and Southern halves with a narrow constriction joining both the halves. The Northern half is larger in size, covering a length of about 12.5 km. from East to West and a breadth of about 7 km. from North to South. The Southern half is smaller in size, covering a length of about 6.5 km. from East to West and a breadth of about 8 km. from North to South. It occupies an area of 100.05 sq. km. between 85^{0}5'18" to 85^{0}17'14" East longitude and 24^{0}19'33" to 24^{0}29'33" North latitude.

The Sanctuary is under the administrative control of Divisional Forest Officer, Wildlife Division, Hazaribag. Notified boundaries of the sanctuary, extending over 20 villages fall under Revenue Thana of Chouparan.

List of the villages of Gautam Buddha Wildlife Sanctuary, Hazaribag

1. Ahri
2. Asnachuan
3. Baniwan tand
4. Bukar
5. Chordaha
6. Danua
7. Dhoria
8. Duuragara
9. Garmorwa
10. Kabilas
11. Kathodumar
12. Khairtanr
13. Mainukhar
14. Mohane tand
15. Morainia
16. Muria
17. Murtiakalan
18. Pathalgara
19. Pathalgarwa
20. Sanjha
21. Sikda
22. Silodhar

The whole Sanctuary area was under private ownership till 1947. This forest suffered maximum damage before they were taken over by the Government after enactment of the Bihar Private Forest Act 1948 and subsequent Land Reform Act in 1950. It was subsequently constituted as protected forest under Indian Forest Act 1927. Again, consequent to the enactment of Wildlife Protection Act 1972, these forests were notified as a Wildlife Sanctuary under section 18 (1) of the said act by the State Government vide its notification No. SO 1485 dated 14.08.1976.

The total area of the sanctuary is 100.05 sq. km. of protected forest. It is highly burdened with rights and concession granted to local villagers. The sanctuary is surrounded by a number of thickly populated villages with cultivation and settlements within, which account for the increasingly undesirable biotic pressure and rendering the area susceptible to degradation. In order to mitigate the biotic pressure from out side sanctuary boundary, the proposal of Eco-fragile zone is under preparation. Besides these there are 20 villages situated inside sanctuary boundary. The area wise largest village is **Ahri,** whereas population wise the largest villages is **Chordaha.**

As per the notification of the sanctuary the protected forest of three different forest divisions namely Gaya Forest Division, Koderma Forest Division and Chatra North Forest Division of Bihar State were taken as Gautam Buddha Wildlife Sanctuary. After creation of Wildlife Division, Hazaribag, the Koderma Forest Division and Chatra North Forest Division are supposed to transfer the forest area to Wildlife Division. The area of Gaya Forest Division went to Bihar State.

After the creation of Jharkhand as a separate state, the area (13833.66 hac.) notified as sanctuary after delineating Gaya Forest Division remain with Bihar state. The area which is expected to be transferred i.e. 2100.82 hac. From Chatra North Forest Division, has not been transferred to Wildlife Division, Hazaribag. So, total effective area of Gautam Buddha Wildlife Sanctuary under control of this division till date covers only 10005.44 hac. Which has been transferred from Koderma Forest Division. The rest are of Chatra North Forest Division are yet to be transferred to Hazaribag Wildlife Division.

2. Approach and Access

The sanctuary is quite easily accessible, as it is well connected by road, rail and air. National Highway No. 2 connecting Howrah to Delhi passes right through the Northern half of the Sanctuary (Plate-III; Fig.-4). The research area has 2 entrances namely Chouparan and Danua check naka. The Chouparan is approximately 41 km. from Koderma, 56 km. from Hazaribag and 70 km. from Gaya.

The nearest railway station is Koderma, which is 41 km. from the Chouparan and is well connected with NH 2. Similarly the nearest airport is Ranchi, which is about 160 km. from Chouparan. Facilities like petrol and diesel filling centers, hotels, hospitals, local transport services such as bus, taxi etc. are available at Chouparan, Hazaribag as well as Koderma town. In addition, communication facilities like telephone, post office etc. are also available.

3. The statement of Significance

The sanctuary occupies a special significance on the Wildlife map of the state, for it consists of uniquely large and compact tract of forest situated on both sides of Howrah to Delhi National Highway No. 2. The sanctuary as of now is suitable for the Sambhar on account of the fact that it's tract, the terrain, the climate and the vegetation make the habitat most suitable for them (Plate-III; Fig.-5). The habitat is shared by some other herbivores like Cheetal, Barking deer, Hare, Wild boar, Fox, Blue bull etc. There is every possibility of occasional visit of carnivores like Hyena, Leopard in the sanctuary area due to its vicinity with Lawalong Wildlife Sanctuary and its continuity with forest area of Gaya Forest Division of Bihar.

Although there is no resident population of elephants, the forests of the sanctuary act as corridor for the internal movement of Indian Elephants by connecting forests of Palamu, Chatra and Koderma and there by help in its migration and out breeding.

The existence of a variety of fruit bearing plants like *Diospyros melanoxylon, Emblica officinalis, Terminalia belerica, Ficus religiosa, Ficus bengalensis* etc. and some almost perennial water pools also attract frequent visits of a large variety of migratory birds. The sanctuary is also valued for its ecological functions such as soil and water conservation. Scientific management of the sanctuary will ensure reduction in erosion and runoff, prolonged flow of water in the streams, reduction in siltation, meeting the irrigation needs of farmers in the down stream etc., as it acts as watershed to feed many small and medium perennial rivers and seasonal streams existing in it.

BACKGROUND INFORMATION AND ATTRIBUTES

I. BOUNDARIES

1. External boundary of Gautam Buddha Wildlife Sanctuary, Hazaribag

The sanctuary is surrounded by Continuous forest belt of Gajhandi Ranges of Koderma forest division and Chatra forest division. It is situated in the Northeast and Western side of Hazaribag District. The boundary line of the sanctuary is given below:

(a) **North:** Villages with protected forest of Barachatti thana, District Gaya.
(b) **South:** Villages with protected forest of Itkhori thana, District Chatra (Forest area falling under the jurisdiction of Koderma Forest Division and Chatra Forest Division).
(c) **East:** Protected Forest area of Itkhori and Chouparan thana falling under the jurisdiction of Koderma Forest Division.
(d) **West:** Forest area Barachatti thana of Gaya district falling under the jurisdiction of Gaya Division, Bihar.

2. Legal boundary

The area within the aforesaid boundaries has been notified as Gautam Buddha Wildlife Sanctuary vide Government of Bihar Notification No. SO 1485 dated 14th September, 1976. As per the notification the total area of the sanctuary should have been 25939.92 hac. or 259.40 sq. km. but due to creation of separate state, 13833.66 hac. or 138.33 sq. km. of sanctuary area remained with Bihar State. The other 2100.82 hac. or 21.00 sq. km. area could not be handed over to Hazaribag Wildlife Division. So, as on today the area of Gautam Buddha Wildlife Sanctuary which is under jurisdiction of Hazaribag Wildlife Division is 100.05 sq. km. of right burdened protected forest and 21.00 sq. km. of sanctuary area is still under jurisdiction of Chatra North Forest Division.

3. Ecological boundary

Based on the migration and excursion patterns of herbivores, which may extend to 5 km. beyond the notified boundaries, the same may be considered as ecological boundaries of the sanctuary for herbivores except elephants, for which it extends very deep into the forests of Koderma, Giridih, Chatra, Ranchi and Palamu Districts. Carnivores like Leopards etc. have been noticed crossing the physical limits of the sanctuary by about 15 km. into neighboring forest. In the year 2006 spurt in cattle killing was reported from Hazaribag Wildlife Sanctuary area. It has been assumed that Leopard or Tiger might have arrived from Palamu Tiger Project through forest or sanctuary area of Gautam Buddha / Lawalong Wildlife Sanctuary. The analysis by Wildlife Institute of India, Dehradun has confirmed the presence of Tiger during that period in Hazaribag Wildlife Sanctuary.

4. Internal boundary

The whole sanctuary consists of one beat i.e. Danua, which is further divided into 4 sub-beats namely Danua, Ahari, Bukar and Garmorwa to be looked after by the respective Forest Guards.

II. ZONE OF INFLUENCE

It is difficult to delimit zone of influence, as there are 22 villages within the sanctuary limits. Due to presence of NH 2 as life line of road transport from New Delhi to Kolkata and presence of transmission line passes through the Northern half of the Sanctuary, the zone of influence is very extensive and scattered. But it can be generalized that the forest area falling within 2 km. radius of each village is under the impact of villagers and therefore, the same is degraded as compared to the rest. Similarly, beyond the sanctuary limits a radius of about 10 km. can be considered as zone of influence in view of the reciprocal influence i.e. research areas influence over people and vice versa.

Considering the above facts and as per the direction of Government of India, it has been decided to send a proposal of Eco-fragile zone around 10 Km. radius of sanctuary area. The process of preparing the proposal of Eco-fragile zone is underway.

TOPOGRAPHY OF THE AREA

Topography (from Greek Topos - place and Grapho - write) is the study of Earth's surface, shape and features. In broader sense, it is concerned with local detail in general, including not only relief but also vegetative and human-made features, and even local history and culture. Survey work is the best form of

Topographical study which was extensively done during the course of excursion and field work.

Topographically most of the area in this region is hilly terrain interspread with grassy plain tracks and river vallies.

The hill system in this area consists of a series of parallel ridges running east and west pierced by about 20 big, moderate river, rivulets and streams which are either perennial or seasonal. There are many spurs and ridges and outlying peaks rising from comparatively open country, which stand in no obvious relation to this system, amongst these may be mentioned the metamorphic rocks which form the water shed between the rivers of the area. The majority of the spurs and ridges bear no definite name. Generally the hills are conspicuous for their irregular form and occurrence. Their contour depends on the nature of rocks of which they are composed.

1. Geology, Rock and Soil

Geologically, two types of formations Archaen type and Gondwanas type exist in this area. Main rock types of Archaen formation are granite and gneiss, whereas sandstone, shale and quartzite constitute Gondwana formation. The Archeans occupy almost whole of the area, while the lower Gondwanas occur in the form of few detached outliners represented by shale, sandstone and boulder bed. The above formations are overlain by the recent alluvium and gravels brought down by the rivers flowing in the area.

Soil is very deep in valleys and plains and is loamy or sandy loam in texture, supporting dense semi deciduous Sal forests interspersed with grasslands, making the sanctuary an ideal habitat for a variety of fauna. On the hills, soils are shallow, while the forest is of miscellaneous type.

2. Terrain

The terrain of the Sanctuary is undulating with low to high hills in the central and western portions. The main hills are Lohawar, Tamasin etc. The altitude of the area varies from 500-600 M above the mean sea level. These hills are separated by valleys, plateaus and stretches of plains. The area is well drained by a series of nalas of streams like Simarkola, Marghati, Goari, Jharna, Sarne Mainuar, Bukar, Mahane etc. The main river is Mahane which originates from Chatra South Forest Division and flows from South to North. Features like gullies, ravines, dams, rock crevices etc are also present at different places of the research area. All these features of the terrain make it an ideal site for a variety of flora and fauna, besides giving it a spectacular landscapes.

3. Climate

The Sanctuary experiences the usual seasons common to a tropical zone i.e. summer, rainy and winter seasons. The summer is characterized by hot and dry winds. The Northwesterly wind in winter causes severe cold conditions in the area.

4. Rainfall

Rains occur mostly during the period of south west monsoon i.e. from June to September. The average yearly rainfall in Hazaribag District is about 1400 mm in the area (Table-1). A chart showing the rainfall pattern during last 4 years is given below:

Table - 1: Rainfall (in mm) details at Gautam Buddha Wildlife Sanctuary, Hazaribag

Month	2007	2008	2009	2010
January	8.0	127.0	53.0	0.0
February	0.0	4.0	0.0	0.0
March	0.0	216.0	0.0	25.0
April	83.0	8.5	0.0	16.5
May	25.0	28.0	177.8	26.8
June	187.0	567.5	94.5	264.7
July	221.5	449.5	379.7	345.1
August	322.5	318.5	217.6	330.0
September	200.4	246.5	365.4	213.0
October	138.0	66.5	65.8	75.3
November	0.0	6.0	29.8	0.0
December	11.5	0.0	21.8	27.5
Total	**1196.90**	**2039.0**	**1405.0**	**1323.9**

5. Temperature

The temperature in the area varies from a minimum of 4^0C in the peak of winter to a maximum of 46^0C c in the summer (Table-2). The temperature pattern during last 4 years is shown below:

Table - 2: Temperature (in centigrade) details at Gautam Buddha Wildlife Sanctuary, Hazaribag

Month	2007		2008		2009		2010	
	Max	Min	Max	Min	Max	Min	Max	Min
January	21.1	6.8	27.6	4.5	29.8	6.0	26.0	3.4
February	25.2	9.4	31.2	3.0	31.2	6.0	30.5	7.0
March	31.7	15.7	35.6	10.5	34.8	10.4	37.6	15.5
April	34.6	19.4	40.0	16.0	40.7	18.0	42.2	19.4
May	45.5	22.3	40.5	19.0	42.2	17.5	37.2	22.7
June	42.0	22.8	38.7	21.4	39.5	18.0	35.3	20.8
July	29.3	22.1	30.8	21.5	31.2	21.0	34.2	21.6
August	28.4	22.2	34.0	21.5	32.8	21.0	31.6	21.0
September	28.7	21.3	32.2	20.5	31.7	21.0	30.8	20.5
October	27.0	16.6	31.3	12.5	29.8	9.0	30.6	15.0
November	24.8	10.1	29.5	7.0	29.3	5.5	29.4	6.5
December	22.6	8.0	28.0	5.5	26.3	4.2	27.0	5.2

6. Humidity

It varies from a maximum of about 100 percent during monsoon to a minimum of around 11 percent, during March/April of the year (Table-3). Humidity pattern in different months during the last 4 years is given below:

Table - 3: Humidity (in percentage) details at Gautam Buddha Wildlife Sanctuary, Hazaribag

Month	2007		2008		2009		2010	
	Max	Min	Max	Min	Max	Min	Max	Min
January	84.9	46.0	98.0	27.0	94.0	21.0	100.0	26.0
February	79.9	29.8	98.0	26.0	73.0	20.0	80.0	25.0
March	49.4	23.4	83.0	24.0	75.0	42.0	89.0	38.0
April	45.5	29.6	56.0	25.0	75.0	11.0	86.0	25.0
May	53.7	37.5	90.0	23.0	95.0	22.0	92.0	26.0
June	72.4	55.2	89.0	24.0	92.0	26.0	92.0	28.0
July	84.0	72.1	90.0	59.0	98.0	62.0	93.0	56.0
August	89.3	84.8	90.0	60.0	100.0	56.0	95.0	60.0
September	85.6	78.0	91.0	55.0	100.0	51.0	95.0	55.0
October	82.9	68.5	92.0	57.0	98.0	65.0	97.0	63.0
November	74.2	62.5	95.0	28.0	94.0	25.0	95.0	30.0
December	81.4	65.3	92.0	34.0	93.0	33.0	92.0	35.0

7. Frost

Incidence of frost has not been recorded but it is known to occur during nights in the peak of winter, i.e. in the months of December and January. Frost cause some damage to the germination and juvenile growths like buds, tips, young leaves etc. The detailed study of impact of frost on flora as well as the behavior of fauna needs to be studied.

8. Natural Water Sources

The sanctuary serves as watershed for a number of streams and rivers. Most of streams are rainfed and seasonal and therefore, dry up in summer. However rivers carry a little bit of water throughout the year and are important source of water for wild animals. The detail list of nalas, streams and rivers are being annexed.

The above water sources besides meeting the needs of the Wildlife also meet the drinking water requirement of cattle (Table-4). In addition, villagers use the streams for domestic and agricultural purposes.

Table - 4: List of Natural Water Bodies (Perennial/Seasonal)

SI. No.	Name of River/ Nala	Location	Remarks
1.	Balwa Nala	Near Ahri, towards S. Border	Seasonal
2.	Bhalupahar River	Near Khairtand, towards Asanchuan	Perennial
3.	Bukar River	Near Bukar, towards Pathalgarhwa and Asanchuan	Perennial
4.	Fatar River	Near Pathargarwa, towards Kathodumar	Perennial
5.	Ghorachaur Nala	Near Garmorwa, towards S.Border	Seasonal
6.	Goari River	Near Chordaha, towards Gaya	Perennial
7.	Guhi Stream	Near Silodhar, towards Danua	Seasonal
8.	Harguma Stream	Near Sikda, towards Murainia	Seasonal
9.	Haribokhar River	Near Muria, towards Gaya	Perennial
10.	Hasail Nala	Near Sikda, towards Pathalgarwa	Seasonal
11.	Jamunia River	Near Asnachuan, towards Khairtand	Perennial
12.	Jarda River	Near Bukar, towords Mainukhar	Perennial

SI. No.	Name of River/ Nala	Location	Remarks
13.	Jharna River	Near Bukar, towards Gajhandi	Perennial
14.	Kadru Nala	Near Murainia, towards Bihar	Seasonal
15.	Khaurawa Ghat River	Near Silodhar, towards Gaya	Perennial
16.	Mainuar River	Near Bukar, towards Pathalgarhwa	Perennial
17.	Marghati River	Near Pathargarwa, towards Mainukhar	Perennial
18.	Mohane River	Near Murania, towards Gaya	Perennial
19.	Panchbadiniya River	Near Asnachuan, towards Bukar	Perennial
20.	Pichri Nala	Near Ahri,	Seasonal
21.	Ramasaran River	Near Muria, towards Chatra	Perennial
22.	Sarne River	Near Bukar, along border of Gaya	Perennial
23.	Simarkola River	Near Pathargarwa, towards Bukar	Perennial
24.	Suhari River	Near Asnachuan, towards Pandria	Perennial
25.	Tari Stream	Near Silodhar, towards Danua	Seasonal

9. Range of Wildlife, Status Distribution and Habitat

(i) Vegetation: Bio geographic classification

From the biogeographical point of view, the area falls in the Oriental Realm (WALLACE). The palaeotropical was divided into three sub-divisions the African, the Indo Malaysian and the Polynesian.

Wallace's formal regions have been modified. It does not differ much from Wallace's original classification, except that it takes into account the distribution of both plant and animal life. The bio-geographical realms together with Wallace's terminology are given below in relation to this area.

Realm	Wallace's terminology	Area included
Indo Malaysian	Oriental	Indian sub continent and South East through CELEBES and SUNDA islands.

(ii) For Types, cover and food for wild animals

According to the classification of the Forest Types of India by Champion and Seth, the forests of the area fall under a broad category of Northern Tropical Dry Deciduous forest, which has been sub-classified as follows:

(a) Dry Peninsular sal (5B/C1c)
(b) Northern dry mixed deciduous forest (5B/C_2).
(c) Dry Bamboo break (5/E_9)

(iii) Dry peninsular Sal

This type of forest is found in valleys or deep inside forest area of Bukar, Asanachuan etc. The crop consists of almost pure sal (*Shorea robusta*) with Asan (*Terminalia tomentosa*) Bija Sal (*Pterocarpus marsupium*), Dhow (*Anogeissus latifolia*) etc. as associated species.

(iv) Northern Dry Mixed Deciduous Forest

This type of forest is present throughout the Research Area except in valleys and depressions. The main species in the upper canopy are Asan, *Terminalia tomentosa, T. chebula, T. belerica,* Salai (*Boswellia serrata*), Kend (*Diospyros melanoxylon*), Piar (*Buchanania latifolia*), *Bauhinia* spp. etc. Small trees and shrubs like *Cleistanthus collinus, Holarrhena antidysenterica, Zizyphus* spp. etc. are found in lower canopy.

(v) Dry Bamboo break

It occurs at scattered places, mixed with dry miscellaneous types of forests. It's main species available here is *Dendrocalamus strictus.* Due to excess and unscientific exploitation, the clump became congested and intangled.

(vi) Other Vegetations

Besides the above three main types of forest, other available types of vegetation deserve a special mention due to their importance from the food and cover point of view. They are as follows:

(a) Lantana Infestation

Lantana weed has infested at many places of the sanctuary especially where degradation has taken place. Though it acts as good cover for small fauna, it does not allow other natural vegetation and grasses to come up in the area and thereby decreases the food base of the wild animals. These areas need greater managerial attention for eradication of this ever proliferating weed and the growth of natural vegetation.

(b) Under story vegetation

Almost in the entire area, except in pure patch of sal, the under storey vegetation consists of *Zizyphus, Nyctanthes arbor-tristis, Wood-fordia fruticosa, Ixora parviflora, Holarrhena antidysentrica* etc. These species act as important food source of herbivores besides grass, especially during dry periods.

(c) Climbers

Common climbers of the area *Bauhinia vahlii, Combretum decandrum, Butea superba, Smilax* spp. etc.

(d) Grasses

Main grasses of the area are *Heteropogon contortus, Chrysopogon* sp., *Dicanthium* etc that occur in open patches and abandoned cultivated lands. Some of these grass species are important sources of food for herbivores. However, wide stretches of grasslands are difficult to find in the research area.

(e) Aquatic vegetation

It is found in the areas where there is a permanent water accumulation. Except a few reservoirs, all the streams and nalas in the research area dry up in summer and therefore not much vegetation is found. However, a few species like *Tamarix dioica* and *Saccharum spontaneum* are found in the moist beds of streams. Among regular aquatics the species such as *Ceratophyllum demersum, Hydrilla verticillata, Cryptocoryne retrospiralis, Vallisneria spiralis* etc. are found.

(f) Sub terrestrial vegetation

The species having roots, rhizomes, bulbs tubers etc. find a mention in this category. These are important sources of food for wild Boars, Bear, Porcupine, Rodents, Monkeys etc. The *Asparagus recemosus, Dendrocalamus strictus* and a few other species having tuberous roots come under this category.

10. Key areas of Conservation importance

(i) **Riparian Zones –** On the banks of the rivers, the riparian zone exists. A wide variety of birds, ungulates, reptiles etc. utilize these area. These area have very wide range of flora and fauna. *Bombax ceiba, Terminalia arjuna, Tamarix* etc are the species abundantly found.

(ii) **Dens, Holes, Sangs –** Dens of Leopard, bears, dholes have been identified in some areas. There are many holes in trees. Snags are also found in the sanctuary area.

REVIEW OF LITERATURE

It should be realized that a 'Flora' will be able to offer critical knowledge of numerous forest products, plants containing vegetable oils, fats and resin, timber, gums, fruits, insecticides, fibres, dyes and medicines or species which may serve for afforestation, for ornamental use, as new green manures, fodder plants or possibly of species withstanding drought or being resistant to fire or inundation suitable for combating erosion and other economic aspects.

The natural vegetation of the District Hazaribag is now facing a severe threat due to increasing urbanization. The present work may help in listing of several endangered, threatened and rare species, moreover, it will create an awareness amongst the people as a whole to protect such species from extinction and to take necessary steps for conservation of our ecosystem and environment. Thus 'Flora' is an essential inventory in this respect and hence a necessity to wide range of users.

The District is floristically rich. The flora of the district is essentially tropophilous though in some parts there is a tendency to xerophily. It usually gives rise to reddish stiff loamy soil, excellently suited for forest growth while kept covered, but bakes to brick like hardness in hot season when denuded.

PREVIOUS WORK

Apart from Hooker's (1872-1897) monumental work and of Prain's (1903) there are several works on flora of Bihar State, e.g., "Plants of Chutianagpur including Jaspur and Sirguja" (Wood, 1902), "A forest flora of Chotanagpur including Gangpur and Santal Parganas" (Haines, 1910), "Botany of Bihar and Orissa" (Haines, 1924), "A Botanical tour in Chotanagpur" (Mukerjee, 1947) and "Supplement to the Botany of Bihar and Orissa" (Mooney, 1950) etc. Other than this, comprehensive account on sedges and grasses flora of this region is available. Some idea about the distribution and occurrence of the sedges and grasses of this area can be gleaned from the study of some recent works of taxonomists from eastern India as in the Flora of Hazaribagh district by Paria and Chattopadhyay, BSI, 2005 and Flora of Palamau by Sharma and Sarkar, BSI, 2002. Ayodhya Singh (1998) in his Ph.D. work under Vinoba Bhave University, Hazaribag gave a comprehensive account of sedges and grasses of Hazaribag District and its neighborhood. Nirbhay Ambasta (2012) in his article on *Cynodon dactylon* (L.) Pers. discussed about its medicinal value. Nirbhay Ambasta and Nawin Kumar Rana (2013) also studied taxonomy of *Chrysopogon aciculatus* (Retz.) Trin. Bhangale and Acharya(2014) studied antiarthritic activity of *Cynodon dactylon* (L.) Pers.

Gordan-Gray (1965) studies Cyperaceae in southern Africa. Comprehensive works were also initiated by Davis (1966), Backer (1968). Primarily exhaustive work on Indian Cyperaceae was carried out by Blatter (1911, 1934) Bor (1938), Caius (1935) and Ambasht (1964). Chavan and Sabnis (1960) reported Cyperaceae from Mount Abu of Rajasthan, India. Clark (1884, 1894, 1898, 1903, 1904, 1908, 1909) worked on Indian Cyperaceae. Cooke (1909) worked on Cyperaceae and reported it in the 'Flora of the Presidency of Bombay'. Among Indian worker D'Almeida and Ramaswamy (1948), Das (1950), Dutta and Mitra (1955) Deshpande and Shah (1968) and Govindarajalu (1966-1982) did their bit towards the study of sedges. Haines (1924), Hall (1973), Haines (1971) Love et al. (1961), Kuekenthal (1909), Kunth (1837) and Linnaeus (1754, 1771). Koyama (1967-1979) was one of the pioneer worker of sedges and owns pathbreaking works in initiating outlines of morphology and classification of Cyperaceae. Similarly Kern (1952, 1961, 1962, 1968, 1974) worked on genera of Cyperaceae of Thailand. Hooker (1854) wrote about Cyperaceae in Himalayan Journal. Hooper (1972) and Guaglianone (1980), Parker (1929), Panigrahi (1965), Ohwi (1965, 1971), Nooteboom (1978), Maheshwari (1950) Malick and Prasad (1968), Saxena (1973), Savile (1979), Sahni (1972), Sabnis (1962, 1967, 1971, 1979), Rao (1974), Rao and Verma (1972, 1975-1977 and 1979) and Raymond (1951, 1955, 1965, 1966) worked on the members of Cyperaceae. Also Singh et al. (2001) BSI, presented an analysis of the sedges and grasses of the then Bihar (Jharkhand) in their exhaustive work. Kumar et al.(2014) studied antibacterial evaluation of *Cyperus rotundus* L. root extracts against respiratory tract pathogens.

Santapau (1951) published, "A Review of Mooney's Supplement to the Botany of Bihar and Orissa", Sanyal (1957) published "Additional notes on the Botany of Bihar and Orissa by H.H. Haines and its Supplement by Dr. Mooney". Other publications are "Recent trends in Flora of Bihar State" (Srivastava, 1959). But due to hilly topography and difficult approachability the District still remains unexplored. Further, large number of plant names used in those books have undergone nomenclatural changes during recent times and needed corrections. Since the days of Hooker, Prain, Haines etc, many neophytes might have naturalized (Srivastava, 1964) and many earlier mentioned species may be on the way of extinction.

In consideration of aforesaid facts and circumstances a comprehensive survey of the sedges and grasses of Gautam Buddha Wildlife Sanctuary, Hazaribag was undertaken.

Sedges have featured in literature since antiquity. The family is well circumscribed and uncontroversial. It was formally described by De Jussieu in 1789; the name is derived from the genus name *Cyperus*, originally from the Greek *Kupeiros*, meaning sedge.

Cynodon dactylon, Desmostachya bipinnata and bamboos have been in use in Hindu ceremonies since long and species of *Cymbopogon* and *Vetiveria* having medicinal properties have been mentioned in the Indian works of 17th and 18th centuries. However, taxonomic treatment of the grasses have been attempted from the 19th century.

Alava (1952) studied spikelet variation in *Zea mays* L. Taxonomical study of the genus *Phalaris* and *Chloris* have been made by Anderson (1961, 1974). Annamalai et al. studied the abnormal stem of *Oryza sativa* L. Ashalatha and Nair (1993) worked on *Brachiaria* in India. *Sporobolus* in Malesia has been worked by Baaijens and Veldkamp (1991) and from sind by Bor (1948). Plants of the Java have been studied by Backer et al. (1968), Barnard (1964) worked on grasses. Bharadwaja (1956) worked on the distribution and origin of the genus *Ischaemum*. Bor (1953-1954 and 1970) worked on *Cymbopogon* and grasses of India. Burkill (1909) also gave notes on *Cymbopogon martinii* Stapf. Chandrasekaran and Daniel (1947) worked on the double digitate inflorescence in *Dactyloctenium aegyptium* Willd. Chatterjee (1947) worked on the wild and cultivated rice. Chinnomani (1975) gave description of Grass Land types and management in India with special reference to *Sehima-Dichanthium*. On grasses depth study has been made by Clayton (1966, 1970, 1972, 1974, 1975, 1979, 1982 and 1989). Cope (1982) also worked on the same in Nasir and Ali's Flora of Pakistan. The grass cover of India has been written by Dabadghao and Shankaranarayan (1973). De Wet et al. (1970, 1984) worked on *Cynodon* and *Eleusine coracana*. Deshpande (1990) wrote *Heteropogon* in India and with Singh (1986), Grasses of Maharastra. Fosberg (1976) gave the status of the name *Chloris barbata* (L.) Swartz. Ambasta and Singh (2010) worked on the taxonomy of *Aristida setacea* Retz. of Hazaribag District.

Gandhi (1968) studied variation in spike number of Star grass (*D. aegyptium* L.). Good (1953) wrote "The Geography of the flowering Plants". Gupta and Jaffer (1982) worked on the rarity and identity of some Indian species of *Cymbopogon*. Henrard (1927-1932) worked on *Aristida*. Monograph on *Digitaria* made by the same in 1950. Judd (1979) wrote Hand book of Tropical Forage grasses. Kaul and Vats (1998) wrote on essential oil composition of *B. pertusa*. A taxonomically interesting deviation in *V. zizanioides* (L.) Nash has been made by Kumar (1963). Lazarides (1976, 1979, 1980 and 1997) worked on the grasses like *Aristida, Eragrostiella, Eragrostis* and *Leptochloa*. Luck (1979) wrote *Setaria* as an important pasture grass. Majumdar et al. (1986) found aflatoxin in *Paspalum scrobiculatum* L. Majumdar (1973) also worked on *Panicum* L. Matthew (1982) gave very beautiful and informative Illustration on the flora of the Tamil Nadu Carnatic. Mehra (1962, 1963) worked on *Dichanthium annulatum* and *Eleusine coracana* (L.) Gaertn. Moulik (1997) worked on the grasses and bamboos of India. Mukherjee (1949, 1954, 1957) worked on *Saccharum* Spp. Murty, 1975 (1972) did taxonomical study of the genus *Apluda* L. Distributional records of grasses have been given by Naithani and Raizada (1977). Cattle poisioning was observed by Nayak of *Digitaria bicornis* have been studied by Pradhan in 1985. Taxonomic and distributional studies in *Leersia* has been made by Pyrah (1969). Prasanna and Pullaiah (1988) studied *Eragrostis* of Andhra Pradesh.

Raizada (1950) reviewed some name changes in Indian grasses. Roy (1979, 1984) studied the genus *Eragrostis* in Rajasthan and wrote Grasses of Madhya Pradesh. Sampath (1963) worked on the taxonomical study of *Oryza* and species relationship. Sharma (1985) worked on *Panicum* of Punjab. Simon (1972) revised the genus *Sacciolepis*. Sivagnanam (1960, 1961) worked on *Eleusine coracana* Gaertn. Skerman and Riveros (1990) worked on Tropical grasses. Sreekumar and Nair (1991) worked on grasses of Kerala. Srivastava and Sinha (2000) worked on Genus *Ischaemum*. Sur (1988 and 2001) revised the genus *Ischaemum* and *Sehima* in India. Tutin et al. (1980) worked on flora of Europe. *Uppuluri* 1975 (1972) worked on taxonomy of the genus *Apluda*

L. Veldkamp (1999, 2001) revised *Chrysopogon* Trin and *Vetiveria* Bory and gave notes on *Chloris*. Watson and Dallwitz (1988) described and Classified Grass genera of the world. Bio systematic study of the genus *Echinochloa* has been made by Yabuno (1966). Yadav and Sardesai (2002) worked on the flora of Kolhapur District. The Biology and utilization of Grasses have been made by Youngner and Mckell in 1972. Patra et al. (2011) study was carried out to assess the nutrient contents of ten grass species like *Cynodon dactylon, Eleusine indica* etc.

An early work on the grasses of Indian sub continent was by Griffith who described grasses of Jheels of Sylhet (now in Bangla Desh). Conspicuous bamboos caught the attention of Kurz (1875) and Gamble (1896). Duthie (1883, 1886, 1888) and Lisboa (1890-1893) collected, studied and described grasses of different regions of India and Sri Lanka, Nineteenth century closed with the publication of monumental work the 'Flora of British India' vol. 7 (Gramineae by Hooker, 1896-1897).

This was followed by the publications of 'Bengal Plants' (Prain, 1903) 'Flora' of Bombay, II (Cooke, 1908), 'Botany of Bihar and Orissa' (Monney, 1950), 'Flora of Madras Presidency' (Gramineae pt. x, Fischer, 1934).

Intensive studies of the grasses were made by Blatter and Mc Cann, their work resulted in the excellent account of the grasses of Bombay (1932). Achariar and Maudaliar (1921) studied the grasses of Madras and the adjoining areas and published the grasses of South India, Jacob contributed to the knowledge of Chittor, Arcot and Travancore; Bor made extensive collections in Assam, Uttar Pradesh and the Himalayas and published grasses of Assam, (Flora of Assam, vol.5. Gramineae, 1940), Common grasses of Uttar Pradesh (1941) and 'Grasses of Burma, Ceylon, India and Pakistan' (1960, reprinted 1973). Stewart (1975) published an account of the grasses of the N.W. India. Raizada, Jain and Bhardwaja (1957) have published 'The grasses of the Upper Gangetic Plain' which is a supplement to the 'Flora of Upper Gangetic Plain' by Duthie. Tewari, Shukla and Panigrahi gave accounts of the grasses of Madhya Pradesh which were precursors to the 'Grasses of Madhya Pradesh' by Roy, (1984). Patunkar (1980) published 'Grasses of Marathawada' (Maharashtra). Several additions and name changes in this family have been published by Shukla and Jain (1978); Uniyal (1986); Singh (1986); Karthikeyan et al. (1989). The latter have brought out a check-list of the Monocots with updated nomenclature. Choudhury studied the grasses of West Bengal. Jain and his co-workers (1961 onwards) have published numerous papers on the Indian grasses including bibliography of the family Gramineae (1961, 1972) and 'Grasses of Bengal, Bihar and Orissa' (1975). They have also revised genera like *Cynodon* (1967). Recently Singh et al. (1979) have published the grasses of Karnataka state and Majumdar (1980) published classification of the Indian grasses.

ECONOMIC IMPORTANCE

The chief importance of sedges and grasses lies in their forming a major natural constituent of wetlands and riverside vegetation, where their densely tangled rhizomes contribute to erosion control and water purification. The consequences once they are eradicated are unfortunately all too easily observed. While on the natural theme, the dense sedge beds that form in swampy regions provide food and shelter for birds, animals and other aquatic life-thus attracting ecotourism. In grasslands, terrestrial game birds feed almost exclusively on the small corms of some *Cyperus* species (Clare Archer, 2005).

Food

Several species are cultivated for their edible grains such as *Coix lachryma-jobi, Eleusine coracana, Hordeum vulgare* (barley), *Oryza sativa* (rice) which forms main food in this region, *Paspalum scrobiculatum, Triticum aestivum* and *Zea mays* (maize) etc.

Several other species such as *Bambusa arundinacea, Dactyloc-tenium aegyptium, Echinochloa colona, Oryza rufipogon* and *Setaria pumila* are also known to provide food at the time of scarcity. Young shoots of the species of bamboos are used as vegetable and are made into pickles.

Fodder

Generally more leafy species are preferred by the stock, though *Chrysopogon aciculatus* with sharp callus awns are also eaten in young stage, before flowerings, by the cattle.

Species considered good for fodder are *Apluda mutica, Cynodon dactylon, Dactyloctenium aegyptium, Digitaria* spp. *Echinochloa colona, Eleusine indica, Hackelochloa granularis, Ischaemum, Rottboellia cochinchinensis* and *Setaria* spp.

Building materials

Species of the Bamboo provides building materials. One layer of these is placed between the wooden squares of the walls and is plastered on both the sides. In some of the states e.g. Mizoram, some parts of Meghalaya etc. most of the houses are entirely built of the bamboos.

Essential oils

Species of *Cymbopogon, Dichanthium* and *Vetiveria* yield oils on steam distillation. Those oils which contain high proportion of Citronella are used for synthesis of certain complicated organic compound e.g. Menthol, whereas those with high Geraniol and low Citronella are used as perfumes in soap and other preparations and extraction of aromatic isolates. In *Vetiveria* the roots contain aromatic oil, these are used for the manufacture of scents and making curtains of 'khas' for use during summer.

Paper industry

Heteropogon contortus, Saccharum spp., yield very good material for paper pulp. *Desmostachya bipinnata, Imperata cylindrica* and *Vetiveria zizanioides* in mixture with the bamboos are being satisfactorily used in the paper industry. *Saccharum officinarum* is used for making inferior quality of wrapping paper.

Lawn grasses

Cynodon dactylon is an ideal lawn grass for lawns, turfs, tennis lawn and Golf's green. Among other grasses *Chrysopogon aciculatus, Imperata cylindrica* and *Oplismenus burmannii* may be used for similar purpose, with regular cutting to check formation of inflorescence. In places having a very heavy rain fall. *Cynodon dactylon* is a very common grass in the lawns and is a chief constituent of the major grass land. *Pennisetum* species with the beautiful coloured inflorescence is an extremely common grass on exposed road sides and in the lawns.

Medicinal grasses

A number of grasses occurring in this region have medicinal properties; some of these have been briefly discussed here:

(i) *Desmostachya bipinnata*: The plant is used in dysentery and menorrhagia. It is also stated to be diuretic.

(ii) *Saccharum officinarum*: Culms and the roots are used as diuretic, cooling agent and aphrodisiac. It is also said to be of use in the intestinal trouble, anaemia, erysepelas and leprosy.
(iii) *Thysanolaena maxima*: Decoction of the roots is used as a mouth wash during fever.

Ornamental grasses

Silvery panicles of *Saccharum* spp. in its natural colour or after dyeing is used for decoration. *Setaria* species with drooping inflorescence is also put to similar use and was once collected in the forest.

Miscellaneous uses

In addition to these, several species are put to various other uses such as:

(1) **Brooms:** Panicles of *Thysanolaena maxima* are made into brooms. This is an important article of commerce in this region.
(2) **Ropes:** Strong ropes are made of *Desmostachya bipinnata.*
(3) **Brewery:** Grains of *Eleusine coracana* and 'involucres' of *Coix lachryma-jobi* are used for this purpose. *Oryza sativa* also has been in use for making good qualities of wines.
(4) **Ornaments and rosary:** The involucres of *Coix lachryma-jobi* are frequently used by Birhors for making necklaces, ear-rings in this area and are also used as beads to make rosary etc. Aromatic leaves of *Cymbopogon* spp. are made into bangles.
(5) **Rodent repellent:** Bristly inflorescence of *Setaria verticillata* are effectively used to protect grains from rodents.
(6) **Hindu rituals:** *Desmostachya bipinnata* and *Cynodon dactylon* are considered to be sacred and are used in many Hindu rituals.
(7) **Soil binder:** Hilly slopes in this region are practically covered all over with *Imperata cylindrica*. Its rhizomes are very effective to check the soil erosion. Similarly root systems of several grasses act as a good soil binder for the surface they cover (Bhimaya et al. 1956).
(8) **Water carriers:** In several places thick culms of the species of bamboos are used for carrying water.

In addition to these uses, bamboos play very important role in the economy of this region. Several articles of daily use such as baskets, mattings, furniture, ladies purses and many other things are made of bamboos.

In the forests of Jharkhand including Gautam Buddha Wildlife Sanctuary, an interesting phenomenon has been observed related with the flowering of *Bambusa* and *Dendrocalamus.* It has been observed the *Bambusa arundinacea* and *Dendrocalamus strictus* flowers periodically, which is related with the rapid increase in population of rodents, these in turn destroy all the available food grains. It is considered inauspicious. Flowering has been reported in 2010 and 2011 from Gautam Buddha Wildlife Sanctuary.

Flowering of the bamboos, its relation with increase of the rodent population, its means of checking etc. need further detailed study.

METHODOLOGY AND PLAN

The present work is based on the results of intensive floristic survey of the aquatic, semi-aquatic and marginal sedges and grasses of Gautam Buddha Wildlife Sanctuary, Hazaribag and the adjoining places conducted during the year 2007-2010.

METHODOLOGY

The work is carried out on the following lines:

For preparing floristic survey of the sedges and grasses, regular field trips were conducted in such a way to cover all the seasons and almost all the parts of the area under investigation on the line suggested by Santapau (1958).

In course of the day's field trip, on the spot notes were entered in the field note book. The data recorded include field number, locality, habit, habitat, height, colour of the flower, association, frequency, local name and other characters which could not be studied from the preserved specimens. On return from day's excursion, all the collections were carefully poisoned in saturated solution of mercuric chloride in rectified spirit and pressed subsequently. However, very slender herbs and delicate flowers were pressed separately in smaller blotters, small herbs were collected entirely, but in case of shrubs or under shrubs, representative portion were taken.

All the specimens were identified with the help of different floras and monographs; those could not be identified were preserved for further study. The taxa of doubtful and unknown identities were matched with their authentic specimens kept at the Central National Herbarium (C.N.H.), Howrah.

Several attempts have been made to note the flowering and fruiting times for most of the taxa.

The specimens were mounted on thick herbarium sheets (42 x 28 cm) with the help of synthetic resin adhesive and threads, separate genus and species cover-sheets were used. All such herbarium specimens after proper labeling were kept in the Herbarium of Department of Botany, Ramgarh College, Ramgarh.

PLAN

The plan followed in the present study is as under:

(i) Systematic Enumeration

In taxonomic treatment, the families have been arranged according to Bentham and Hooker's (1862-83) system of classification.

(ii) Keys

On the basis of exhaustive studies, simple keys have been designed to assist in the identification of local taxa. All the keys are dichotomous and are based on macroscopic/microscopic characters. A general key to the genera of each family and key to the species of each genus have also been provided.

The genera within the family and species within the genus are arranged in alphabetical sequences.

(iii) Nomenclature

Nomenclature of the taxa of the area have been made up to date in accordance with the International Code of Nomenclature for algae, fungi, and plants (I.C.N.). For each taxon, the latest of correct name with full reference of author(s), basionym, if any and synonyms to the names accepted in 'Flora of British India' and 'The Botany of Bihar and Orissa' and references to the latest monographs and taxonomic revisions have been given.

(v) Descriptions

An attempt has been made to give a concise description of all the plants included, which is followed by field notes, local names have been given in all those cases where they could be ascertained properly. Flowering and fruiting times have been critically observed. In the citation of specimens, collector's name is followed by collection number. Uses, if any for medicinal and other purposes have also been mentioned.

In the systematic part, key to the genera and key to the species are provided. Genera and species are alphabetically arranged within the family/genus as the case may be.

For the correct names several works of Raizada (1948, 1959 and 1966), Bennet (1987), Karthikeyan et al. (1989), Bridson and Smith (1991), Brummitt and Powell (1996), Featherly (1973), Lawrence et al. (1968), McNeill et al. (2006) and Stafleu and Mennega (1992-2000) are consulted. Among the latest works, Flora of India, edited by Sharma et al. (1993) vol. 1, Sharma and Balakrishnan (1993) vol. 2, Sharma and Sanjappa (1993) vol. 3 and Hajra et al. (1995) vol. 12 and 13. Flora of West Bengal by Bhattacharyya et al. (1997) and Flora of West Champaran by Bhattacharyya and Sarkar (1998) are also consulted. Nomenclature has been brought up to date in accordance with I.C.B.N. as far as practicable. Brief descriptions of species with flowering and fruiting time is given with notes on nomenclature, habit, economic aspects (Watt, 1889; Chopra et al., 1956 and 1969; Trivedi et al., 1985 and Ambasta, 1992) including local uses if any, local names of the plant together with Hindi.

(vi) Present Work

The study of Cyperaceae and Poaceae of Gautam Buddha Wildlife Sanctuary deals with the general consideration of the flora which includes introduction, materials and general account dealing with area, topography, geology and soils, climate, vegetation and biotic influences over vegetation. This is followed by a statistical analysis of the members of the two families.

GLUMACEAE

Key to the Families

1a. Mostly perennial with 3-quetrous stem and 3-farious leaves
.... 1. Cyperaceae

1b. Annual or perennial with terete stem and distichous leaves
.... 2. Poaceae

1. CYPERACEAE

Key to the Genera

1a. Flowers unisexual:
Pistillate flowers enclosed in a sac-likeorgan (utricle) ...... 2. *Carex*

1b. Flowers bisexual [sometimes also staminate in distal glume (s) of *Kyllinga*]:

2a. Style continuing down to ovary without demarcated border, hence nut neither crowned by, nor jointed with style-base:

3a. Glumes spirally imbricate. Hypogynous scales and/or bristles present (exceptions: *Schoenoplectus*):

4a. Hypogynous scales 3, alternating with 3 bristles. Glumes pilose outside ...... 5. *Fuirena*

4b. Hypogynous scales absent, bristles 0-6. Glumes glabrous ...... 9. *Schoenoplectus*

3b. Glumes distichously arranged. Hypogynous scales and bristles absent:

5a. Rachilla articulated, hence spikelets falling in entirely:

6a. Style 2-fid. Nuts bilaterally flattened with one angle facing rachilla ...... 6. *Kyllinga*

6b. Style 3-fid. Nuts trigonous,

with one side facing rachilla ...... 7. *Mariscus*

5b. Rachilla not articulated, persistent, hence glumes falling apart from rachilla:

7a. Nuts trigonous or dorsiventrally flattened with one side facing the rachilla ...... 3. *Cyperus*

7b. Nuts bilaterally flattened with one angle facing the rachilla ...... 8. *Pycreus*

2b. Style jointed with the ovary; style-base dilated or spongy-thickened, clearly demarcated:

8a. Nut crowned by persistent style-base ...... 1. *Bulbostylis*

8b. Nut not crowned by persistent style-base ...... 4. *Fimbristylis*

1. BULBOSTYLIS Kunth nom. cons.

Key to the Species

1a. Inflorescence capitate ...... 1. *Bulbostylis barbata*

1b. Inflorescence anthelate ...... 2. *Bulbostylis densa*

1. Bulbostylis barbata (Rottb.) Kunth ex Clarke in Hook. f., Fl. Brit. India 6: 651. 1893; Prain, Bengal Pl. 2: 1156. 1903 (Rep. ed. 2: 870. 1963); Haines, Bot. Bihar & Orissa pt. 5: 923.1924; Blatter and McCann in J. Bombay Nat. Hist. Soc. 37: 768. 1935; Kern in Reinwardtia 6: 51. 1961: Koyama in Bot. Mag. Tokyo 93: 341. 1980; Singh et al., Fl. Bihar 559. 2001; Paria & Chattopadhyay, Fl. Hazaribagh District 2: 980. 2005; *Scripus barbatus* Rottb., Progr. 27. 1772; *Fimbristylis barbatta* (Rottb.) Benth., Fl. Austral. 7: 321. 1878; Koyama in J. Fac. Sc. Univ. Tokyo Sect. 3. 8: 104. 1961.

Local name: Masa.

A small tufted annual. Stem 5-25 cm high, slender, glabrous. Leaves as long as or shorter than stem, capillary; sheath membranous, pilose. Spikelets 3-7 mm long, oblong-lanceolate, few-flowered, brown, crowded in a terminal head; bracts 3, filiform. Glumes 1.2-2.5 mm long laterally compressed, broadly ovate when unfolded, loosely imbricate, keel strong and ending in a mucro. Stamen 1. Nut 0.5 mm long, broadly obovoid, trigonous, smooth, straw-coloured.

Ecology: Frequent on sandy river-beds and forest edges.
Fl. & Fr.: July-Oct.
Distribution: Throughout India. Widespread in tropical areas of the world.
Specimens examined: Hathia baba, 1001; Danua, 2021.
Uses: Plant is boiled in water and used in dysentery.

2. Bulbostylis densa (Wallich) Hand.- Mazz. in G. Karst. & Schenk, Vegetationsb. 20.7: 16, 1930; Verma & Chandra in Rec. Bot. Surv. India 21(2) 226. 1981; Singh et al., Fl. Bihar 559: 2001; *Scirpus densus* Wallich in Roxb. Fl. Ind. (Carey & Wallich ed.) 1: 231. 1820. *Bulbostylis capillaris,* Kunth. var. *trifida* (Nees) C.B. Clarke in Hook. f., Fl. Brit. India 6: 652. 1893; Haines, Bot. Bihar & Orissa pt. 5: 924. 1924; *Isolepis trifida* Nees in Wight, Contr. Bot. India 108. 1834.

Annuals, 5-40 cm tall. Leaves shorter than the stem, ca 0.2 mm broad; sheaths white hairy at the mouth. Anthela lax, simple or compound bracts setaceous, awned; rays one to several, ascending. Spikelets usually solitary, sometimes a few clustered, oblong-ovoid, polygonal, 3-6x1.8-2.3 mm. Glumes laxly imbricating, brown, 1.5-2 mm long, acute or muticous. Nuts stramineous or ultimately brown, oblong- obovoid, 0.7-0.9 x 0.5-0.8 mm, verruculose, transversely lineolate; surface cells transversely oblong.

Ecology: Common on sandy river-beds and moist sandy places.
Fl. & Fr.: Sept.-Dec.
Distribution: Throughout India.
Specimens examined: Chordaha, 1051; Danua, 1071.

2. CARREX L.

Carex cruciata Wahlenb., Vet. Akad. Handl. Stockh. 24: 149. 1803; Clark in Hook. f., Fl. Brit. India 6: 715. 1894; Prain, Bengal Pl. 2: 1130. 1903; Kukenth. in Engl., Pfl. - reich IV-20, Heft 38: 265. 1909: Haines, Bot. Bihar & Orissa pt. 5: 934. 1924. p.p.: Kern & Nooteb. in Steenis, Fl. Males. ser. 1, 9: 121. 1979; Singh et al., Fl. Bihar 559: 2001; Paria & Chattopadhyay, Fl. Hazaribagh District 2: 981. 2005. *C. condensata* Nees in Wight Contr. Bot. 123. 1834: Clark in Hook. f., Fl. Brit. India 6: 716. 1894.

Perennials, up to 1.5 m tall. Leaves exceeding the stem, 4-13 mm broad, coriaceous; sheaths reddish brown. Panicles 15-45 (-60) cm long; partial panicles (3-) 5-11, lower on long exserted peduncles; rachis hispidulous; bracts equaling to much exceeding the panicle. Spikelets numerous, androgynous, 6-15 mm long; cladoprophylls utriculiform. Glumes stramineous or brownish, with purplish brown streaks, ovate, upper female acute or mucronulate, lower with up to 1 mm long awn. Style swollen at the base; stigmas 3. Utricles divaricate, inflated, ellipsoid or ovate-oblong, subtrigonous, 3-4.5 x 1.2-1.7 mm, abruptly beaked, few-nerved, glabrous or hispidulous; beak 2-dentate. Nuts blackish brown, ellipsoid, trigonous-triquetrous, 1.5-2.2 mm long.

Ecology: Common along margin of forests, hanging from hill-slops, etc.

Fl. & Fr.: July-Aug.

Distribution: Himalayan region, Eastern India and E. peninsula. Bhutan, Nepal, Lower Mayanmar, Indo-China, Japan, Malaysia.

Specimens examined: Sanjha, 2020; Garmorwa, 2001.

3. CYPERUS L.

Key to the Species

1a. Nuts corky-thickened on the angles:
Inflorescence anthelate; style deeply 3-fid
1b. Nuts not corky-thickened on the angles: ...... 10. *C. platystylis*
2a. Stigmas 2; nuts lenticular:
3a. Glumes mucronate or mucronulate:
Inflorescence of a few
glomerules collected into a

head; spikelets 3-5 mm long; lower glumes spiral, upper distichous ...... 13. *C. pygmaeus*

3b. Glumes acute or muticous: Spikelets more or less erect, in dense clusters; anthers 0.4 mm long ...... 11. *C. polystachyos*

2b. Stigmas 3; nuts trigonous:

4a. Rachilla of the spikelets caducous, the spikelets thus falling off as a whole along with the persistent glumes: ...... 5. *C. dubius*

4b. Rachilla of the spikelets persistent, the glumes acropetally caducous:

5a. Stamen 1(-2); nuts linear or narrowly oblong with almost parallel sides ...... 2. *C. castaneus*

5b. Stamens 1-3; nuts obovoid-oblong or subglobose:

5b. Stamens 1-3; nuts obovoid-oblong or subglobose:

6a. Inflorescence capitate: Glumes broadly obovate-suborbicular, 0.5-0.8 mm long ...... 3. *C. difformis*

6b. Inflorescence anthelate with distinct rays:

7a. Spikelets in well elongated spikes; the rachis several times longer than the spikelets:

8a. Glumes orbicular or broadly ovate about as broad as long ......7. *C. iria*

8b. Glumes linear-lanceolate, ovate or oblong, about half or less as broad as long:

9a. Rachis hispidulous: Anthela simple or

subcompound; spikelets 10-35 mm long; nuts ca 1.5 mm long ...... 12. *C. procerus*

9b. Rachis glabrous: Spikes penicillate, 2-10 mm broad, with ascending spikelets

7b. Spikelets digitate or in very short spikes, the rachis obscure or much shorter than to about as long as the spikelets:

10a. Spikelets digitate, 3-20 mm long:

11a. Glumes mucronate: Leaves 1.5-4 mm broad; Bracts all shorter than or scarcely equaling the anthela ...... 6. *C. halpan*

11b. Glumes muticous or mucronulate: Rachilla visible between the widely spreading glumes (prominently so in the mature spikelets); glumes 0.7-1 mm long ...... 15. *C. tenuispica*

10b. Spikelets shortly spicate, 5-60 mm long:

12a. Stems stout, 4-8 mm thick in the middle; leaves absent or solitary: Lower bracts much exceeding the anthela ......9. *C. pangorei*

12b. Stems slender or stout, 1-5 mm thick in the

middle; leaves 2-several:

13a. Spikelets acicular, ca 1 mm broad, glumes 1.5-2 mm long ...... 4. *C. distans*

13b. Spikelets compressed, 1-3 mm broad; glumes 2-4.5 mm long:

14a. Stem base enclosed in a hard irregularly splitting coat; involucral bracts distinctly separated from one another or only the lowermost obvious ...... 1. *C. bulbosus*

14b. Stem base covered with membranaceous scales when young, the scales soon disintegrating into fibres; involucral bracts close together or only the lowermost a little distant:

Tubers without grey tomentum; Spikelets 10-35 mm long, 10-40-flowered;

glumes with 5-7 nerves, the lateral nerves becoming much less prominent away from the keel, usually deep brown, rarely yellowish ...... 14. *C. rotundus*

1. Cyperus bulbosus Vahl, Enum. Pl. 2: 342. 1805; C. B. Clarke in Hook f., Fl. Brit. India 6: 611. 1893; Kwen in Steenis, Fl. Males. Ser. I. 7(3): 605. 1974. Singh et al., Fl. Madhya Pradesh 3: 258. 2001; Singh et al., Fl. Bihar 562. 2001.

Perennials, 10-30 cm high. Rhizomes bearing long slender stolons. Stems solitary, bulbously swollen at base due to a cover of several stiff and irregularly splitting chestnut brown or blackish scales. Leaves usually shorter than the stem. Anthela usually simple, with up to 2 cm long divaricate rays, sometimes subcapitately contracted; bracts 4-8, foliaceous. Spikes cylindrical, 1-3 cm long. Spikelets linear- ellipsoid, bearing a solitary nut. Glumes stramineous, oblong-lanceolate, subacute, 9-11-nerved, with a green keel. Nuts dark brown or blackish, ellipsoid- obovoid, ca 1.5 x 0.5 mm.

Ecology: Found in stagnant water.
Fl. & Fr.: July-Oct.
Distribution: Throughout India
Specimens examined: Bukar, 8011; Mohane tand, 8082.

2. Cyperus castaneus Willd., Sp. Pl. 1: 278. 1797; Clarke in Hook. f., Fl. Brit. India 6: 598. 1893; Haines, Bot. Bihar & Orissa pt. 5: 894. 1924; Blatter and McCann in J. Bombay Nat. Hist. Soc. 37: 258. 1935; Kuekenthal in Engler, Pfl-reich 101: 264. 1936; Kern in Reinwardtia 2:117. 1952 & in van Steenis, FI. Malesiana ser. 1. 7: 630. 1974; Koyama in Garden's Bull. Singapore 30: 143.1977; Singh et al., Fl. Bihar 562. 2001.

Annuals, tufted, 3-15 cm high. Leaves filiform, usually about as long as the stem. Anthela usually simple, sometimes compound, or sometimes reduced to a head; bracts filiform, up to 10 cm long. Spikelets 3-30 in a cluster, chestnut brown (or sometimes pale), linear, 5-20 mm long, ca 1.5 mm broad. Glumes oblong-spathulate, plicate 1-1.5mm long with a 3-nerved keel excurrent into a greenish yellow recurved mucro. Stamens 1 (-2). Stigmas 3. Nuts dark red, linear or narrowly oblong with almost parallel sides, ca 1 mm long.

Ecology: Found in grassy moist places.
Fl. & Fr.: Aug.-December
Distribution: Pantropic.
Specimens examined: Pathalgarwa, 1069; Asnachuan, 2011.

3. Cyperus difformis L., Cent. Pl. 2: 6. 1756; Clarke in Hook. f., Fl. Brit. India 6: 599. 1893; Prain, Bengal Pl. 2: 1142. 1903 (Rep. ed 2: 859. 1963); Haines, Bot. Bihar & Orissa pt. 5: 893. 1924; Blatter and McCann in J. Bombay Nat. Hist. Soc. 37: 259. 1935; Kuekenthal in Engler, Pfl.-reich 101: 237. 1963; Koyama in Quart. J. Taiwan Mus. 14: 174. 1961 & in Gardens' Bull. Singapore 30: 143. 1977; Kern in Reinwardtia 6: 58. 1961 & in van Steenis, Fl. Malesiana ser. 1. 7: 629. 1974; Singh et al., Fl. Bihar 563. 2001; Paria & Chattopadhyay, Fl. Hazaribagh District 2: 988. 2005.

Tufted annual. Stems 10-40 cm high, glabrous, 3-quetrous. Leaves flaccid, often shorter than the stem, linear, acuminate, 3-4 mm broad. Inflorescence simple or compound, sometimes reduced to a head; bracts 2-3, the lowest up to 25 cm long, leaf-like. Spikelets many, brown, 3-5 x 1-1.25 mm; rachilla slender, not winged. Glumes 0.5 mm long or less, broadly obovate-oblong, obtuse, concave, imbricate, 3-nerved. Stamens 1, rarely 2. Nut 0.25 mm long, broadly ellipsoid, pale brown.

Ecology: Common in Sandy river beds, wet grounds, rice fields etc.
Fl. & Fr.: Aug.-Jan.
Distribution: Throughout India.
Specimens examined: Asnachuan, 1104; Chamargadda, 1037.

4. Cyperus distans L. f., Suppl. 103. 1781; Clarke in Hook. f., Fl. Brit. India 6: 607. 1893; Prain, Bengal Pl. 2: 1143. 1903 (Rep. ed. 2: 861. 1963); Haines, Bot. Bihar & Orissa pt. 5: 898. 1924; Blatter and McCann in J. Bombay Nat. Hist. Soc. 37: 266. 1935; Kuekenthal in Engler, Pfl.-reich 101:

137. 1936; Koyama in Quart J. Taiwan Mus. 14: 169. 1936 & in Gardens' Bull. Singapore 30: 135. 1977; Kern in Reinwardtia 6: 53. 1961 & in van Steenis, fl. Malesiana ser 1. 7: 610. 1974; Singh et al., Fl. Bihar 564. 2001; Paria & Chattopadhyay, Fl. Hazaribagh District 2: 989. 2005.

Perennials, 0.2-1 m high. Rhizomes short, bearing short stout stolons. Stems stout, triquetrous, swollen at base. Leaves 3-10 mm broad. Anthela usually decompound, sometimes compound or supradecompound, 10-30 cm across. Spikes lax, pyramidal. Spikelets young ascending, older divaricate or deflexed, linear or somewhat acicular, 6-60 x 1 mm; rachilla flexuous, the internodes ca 1 mm apart; rachilla wings hyaline, lanceolate, soon caducous. Glumes rather remote, appressed, usually reddish brown, sometimes yellowish, elliptic, 1.5-2 mm long, muticous, scarcely keeled, slightly inrolled on margins. Stamens 3. Stigmas 3. Nuts deep brown, ellipsoid, trigonous, 1.3-1.6 mm long, apiculate.

Ecology: Common in river banks, grassy road sides, wet rice-fields etc.
Fl. & Fr.: Aug.-Nov.
Distribution: Throughout India. Pantropic.
Specimen examined: Danua, 1047.

5. Cyperus dubius Rottb., Descr. Icon. Rar. Pl. 20. t. 4. f. 5. 1773; Verma & Chandra in Rec. Bot. Surv. India 21(2): 240. 1981; Singh et al., Fl. Madhya Pradesh 3: 265. 2001. *Mariscus dregeanus* Kunth, Enum. Pl. 2: 120. 1837; C. B. Clarke in Hook. f., Fl. Brit. India 6: 620. 1893; Haines, Bot. Bihar & Orissa pt. 5: 908. 1924. *M. dubius* Fischer in Gamble, Fl. Madras 3: 1644. 1931.

Perennials, 10-50 cm high. Stems slender, tufted, triquetrous, thickened at base by turgid chestnut brown sheaths. Leaves flaccid, 2-3 mm broad. Inflorescence a solitary (rarely 2-4 together), terminal ovoid, dense, 7-15 mm broad head; bracts 3-5 or more, foliaceous, upto 35 cm long. Spikelets numerous congested into the head, ovoid, 5-8 mm long, bearing 2-6 nuts; rachilla wings elliptic. Glumes pale brown, ovate- triangular, muticous. Stamens 2-3. Stigmas 3. Nuts black, oblong-subovoid, trigonous, slightly shorter than the glume.

Ecology: Common in water-lodged area.
Fl. & Fr.: June-Nov.
Distribution: Almost throughout India.
Specimens examined: Dhoria, 8009; Kabilas, 8032.

6. Cyperus halpan L., Sp. Pl. 45. 1753 ('haspan'); Clarke in Hook. f., Fl. Brit. India 6: 600. 1893; Prain, Bengal Pl. 2: 1142. 1903; Haines, Bot. Bihar & Orissa pt. 5: 894. 1924; Kern in Steenis, Fl. Males. ser. 1. 7: 672. 1974; Koyama in Dassan. & Fosb., Rev. Handb. F. Ceylon 5: 203. 1985; Singh et al., Fl. Bihar 564. 2001; Paria & Chattopadhyay, Fl. Hazaribagh District 2: 990. 2005.

Perennials or rarely annuals, 8-70 cm high. Rhizomes slender, short or up to 5 cm long, creeping. Stems slender, compressed-trigonous, sometimes almost absent in depauperated plants. Leaves 1.5-4 mm broad, sometimes absent. Anthela compound or decompound, lax; bracts -3, foliaceous. Spikelets digitate, (2-) 3-6 (-13) together, linear-lanceolate, 2.5-13 x 1-1.5 mm. Glumes ascending and appressed to the rachilla, even in fruits, so that the rachilla is invisible or scarcely visible, yellowish red, ovate-oblong. 1-1.7 mm long, mucronulate, inrolled on margins. Stamens 1 (3-). Style short, 3-fid. Nuts yellowish, broadly obovoid, trigonous, 0.4-0.5 mm long, apiculate, shortly stipitate.

Ecology: Common in wet places, rice-field, grasslands, etc.
Fl. & Fr.: June-Oct.
Distribution: Throughout India. Tropical and subtropical regions of whole world.
Specimens examined: Khairtand, 2016; Siarkoni, 2007.
Uses: Plant is used as fodder.

7. Cyperus iria L. Sp. Pl. 45. 1753; Haines, Bot. Bihar & Orissa pt. 5: 895. 1924; FBI 6: 606; Kuekenthal in Pflanzenr. 101: 150. 1935; BP 2: 860; Kern in Backer & Bakh. f. Fl. Java 3: 479. 1968; Singh et al., Fl. Bihar 565. 2001; Paria & Chattopadhyay, Fl. Hazaribagh District 2: 991. 2005.

Annual. Stems tufted, slender or slightly stout, triquetrous, 15-50 cm tall and 2-3 mm thick. Leaves basal, flat or channeled acuminate, scabrous on margins in upper parts, shorter than to slightly surpassing the stems, 3-6 mm wide. Inflorescence simple or compound, loose, 5-20 cm long. Involucral bracts 3-5, the larger 1-3 overtopping the inflorescence, to 40 cm long. Primary rays 3-5, unequal, to 10 cm long, the larger ones usually branched; secondary rays very short. Spikes oblong-ovoid usually elongate, 1-5 cm long, bearing 5-25 spikelets. Spikelets oblong-linear, compressed, 3-10 mm long and 1.5-2 mm wide 6-25-flowered, rachilla persistent, wingless. Glumes orbicular or broadly ovate, usually broader than long, rounded to emarginate at apex, shortly mucronulate, 1.2-1.5 mm long and 1-1.2 mm wide with green, arched, 3-5-nerved sharp keel, golden,-fulvous, nerveless sides and whitish hyaline margins towards top. Stamens 2-3. Style short; stigmas 3. Nut obovoid-ellipsoid, triquetrous with concave sides, broadly stipitate, shining, dark-brown to black, 1-1.5 mm long and 0.5-0.7 mm wide (Plate-XXVII; Fig.-29A and B).

Ecology: Very common in open wet places, river-banks, wet rice-fields, etc.
Fl. & Fr.: Aug.-Dec.
Distribution: Throughout India. Widespread in tropical Asia, extending northwards to Iran, Afghanistan, China and Japan.
Southwards to Malaysia and Australia and westwards to tropical E. Africa; introduced in S.E. United states and West Indies.
Specimens examined: Garmorwa, 1057; Sanjha, 1005; Siarkoni, 1068.
Uses: Plant is used as fodder. Stems are woven into mats; also used as astringent, stomachic and tonic.

8. Cyperus nutans Vahl, Enum. 2: 363. 1836; Clarke in Hook. f., Fl. Brit India 6: 607. 1893; Prain, Bengal Pl. 2: 1144. 1903; (Rep. ed. 2: 861. 1963); Haines, Bot. Bihar & Orissa pt. 5: 898. 1924; Blatter and McCann in J. Bombay Nat. Hist. Soc. 37: 266. 1935; Kuekenthal in Engler. Pfl.-reich 101: 144. 1936; Koyama in J. Taiwan Mus. 14: 168. 1961 & in van Steenis, Fl. Malesiana ser. 1. 7: 609. 1974; Singh et al., Fl. Bihar 565. 2001.

Perennials, 0.5-1 m high. Rhizomes short, bearing short stolons. Stems stout, trigonous below, triquetrous above. Leaves 6-12 mm broad, coriaceous. Anthela lax, compound or decompound; bracts 4-7, foliaceous, up to 60 cm long; rays obliquely spreading, compressed. Spikes dense, penicilliform, often somewhat nodding, 3-4.5x0.2-1 cm, bearing 15-25 spikelets. Spikelets linear-oblong, 5-14x2 mm, 8-16 – flowered; rachilla wings hyaline, lanceolate; internodes ca 0.7 mm long. Glumes elliptic, 2-2.5 x 1 mm, margins broadly scarious in the upper half, incised at the tip and mucronulate by the excurrent tip, 7-nerved; keel grayish green, sides pale fuscous. Stamens 3. Styles 3-fid almost to the base. Nuts brown, narrowly obovoid-oblong, triquetrous, ca 1.5x0.5 mm, apiculate.

1a. Spikelets 8-16 flowered; internodes of rachilla ca 0.7 mm long; nuts ca 1.5 x 0.5 mm ...... var. *nutans*

1b. Spikelets less than 8-flowered; internodes or rachilla ca 0.5mm long; nuts 1.1-1.2 x 0.6-0.7 mm ...... var. *eleusinoides*

var. **eleusinoides** (Kunth) Haines, Bot. Bihar & Orissa pt. 5: 898. 1924. *Cyperus eleusinoides* Kunth, Enum. Pl. 2: 239. 1837; C.B. Clarke in Hook. f., Fl. Brit. India 6: 608. 1893. *C. nutans* subsp. *eleusinoides* Koyama in Gard. Bull. 30: 136. 1977.

Ecology: Common along river, wet places, etc.
Fl. & Fr.: July-Dec.
Specimens examined: Chordaha, 1072; Danua, 1073.
Distribution: Throughout India. Tropical and Sub-tropical regions of the world.

9. Cyperus pangorei Rottb., Descr. Icon. Rar. Pl. 31. t. 7. f. 3. 1977. non *C. corymbosus* var. *pangorei* C.B. Clarke in Hook. f., Fl. Brit. India 6: 612. 1893; Singh et al., Fl. Bihar 566. 2001. *C. tegetum* Roxb., Fl. Ind. 1: 208. 1832; C. B. Clarke in Hook f., Fl. Brit. India 6: 613. 1893; Haines, Bot. Bihar & Orissa pt. 5: 900. 1924.

Perennial with short, stout rhizome. Stems 45-100 cm high, 3- angular. Leaves often much reduced sometimes 25 cm or more long, 4-8 mm broad, linear, acuminate. Anthela compound or decompound; rays 6-10, each of which bearing 4-10 slender spikelets, brackets often 15-30 cm long, erecto-patent, linear, acuminate. Spikelets 0.8-1.8 x 0.25 cm, 10-30-flowered, brown or reddish-brown, linear; rachilla winged. Glumes 3x1.25 mm, oblong or oblong-ellipsoid, imbricate; keel rounded, green, with 5 nerves and scarious margins. Stamens 3. Nut obovoid-oblong, yellowish-brown.

Ecology: Occurs chiefly on river banks
Fl. & Fr.: Aug.-Nov.
Distribution: Throughout India.
Specimens examined: Silodhar, 2005; Murtiakalan, 2017.

10. Cyperus platystylis R. Br., Fl. Nov. Holl. 215. 1810; Clarke in Hook. f., Fl. Brit. India 6: 598. 1893; Prain, Bengal Pl. 2: 1141. 1903; Haines, Bot. Bihar & Orissa pt. 5: 893. 1924; Singh et al., Fl. Bihar 566. 2001; Paria & Chattopadhyay, Fl. Hazaribagh District 2: 994. 2005.

Perennials, aquatic. Rhizomes short. Stems triquetrous, 0.1-1 m long, smooth or scabrid at the tip. Leaves grayish green, 7-12 (-20) mm broad, coriaceous, transversely septate, scabrid on margins and midrib beneath; sheaths strongly keeled. Anthela decomound, 6-30 cm across; bracts 5-8, foliaceous, up to 80 cm long. Spikelets in clusters of 3-8, ovate, lanceolate, 6-15 (-20) x 2.5-3 mm. Glumes glossy yellowish brown, broadly ovate, 2-2.5 mm long, mucronate, membranaceous, cellular-reticulate. Stamens 3. Stigmas 3. Nuts grayish brown, ellipsoid, trigonous, ventrally concave, dorsally convex with raised angles, 1.7-2 mm long, apiculate, the angles in mature nuts corky thickened.

Ecology: Along ponds, wet places, etc.
Fl. & Fr.: May-Nov.
Distribution: Bihar, Jharkhand, West Bengal, Bangladesh, Sri Lanka to Taiwan and Malaysia eastwards to northern and eastern Australia.
Specimens examined*:* Bukar, 2010; Khairtand, 2014.

11. Cyperus polystachyos Rottb., Descr. Pl. Rar. 21. 1772 & Descr. Icon. Rar. Pl. 39. t. 21. f. 1. 1773; Haines, Bot. Bihar & Orissa pt. 5: 903. 1924; Singh et al., Fl. Madhya Pradesh 3: 278. 2001. *Pycreus polystachyos* P. Beauv., Fl. Oware 2: 48. t. 86. f. 2. 1807; C. B. Clarke in Hook. f., Fl. Brit. India 6: 592. 1893.

Annuals or perennials, 10-70 cm high. Stem compressed-trigonous. Leaves grayish green, 2-4 mm broad. Anthela compound, evolute or capitately contracted; bracts 3-6, foliaceous, upto 20 cm long. Spikelets numerous, congested, linear, 8-25x1.5 mm; rachilla narrowly winged. Glumes stramineous or pale ferrugineous (rarely castaneous), ovate- lanceolate, 1.5-2 mm long, acute, membranaceous. Stamens 1-2. Styles deeply 2-fid. Nuts brown or ultimately glossy black, linear-oblong, lenticular, with almost truncate shoulders, 1-1.2 mm long, apiculate.

Ecology: Common in water-lodged areas.
Fl. & Fr.: Aug.-Dec.
Distribution: Throughout India.
Specimens examined: Kabilas, 8012; Manukhar, 8075.

12. Cyperus procerus Rottb., Descr. & Ic. Rar. Nov. Pl. 29. t. 5. f. 3. 1773; Clarke in Hook f., Fl. Brit. Ind. 6: 610. 1893; Prain, Bengal Pl. 2: 1143.

1903; Haines, Bot. Bihar & Orissa pt. 5: 902.1924; Singh et al., Fl. Bihar 566. 2001; Paria & Chattopadhyay, Fl. Hazaribagh District 2: 995. 2005.

Perennials, tufted, up to 1.2 m high. Rhizomes stoloniferous. Stems stout, triquetrous. Leaves longer or shorter than the stem, 9-15 mm broad, coriaceous, canaliculated. Anthela simple or subcompound, 10-15 cm across; bracts foliaceous, exceeding the anthela; rays rigid, 2-10 cm long, spreading. Spikes diverging, shortly pedunculate; rachis hyaline on margins, sparsely hispidulous. Spikelets oblong or broadly linear, 10-35 mm long, 10-50 - flowered, Glumes loosely imbricating, reddish brown, ovate or elliptic, 2.5-3 mm long, mucronulate, 5-7 nerved, hyaline on margins. Stamens 3. Style deeply 3-fid. Nuts blackish brown, obovoid or ellipsoid, trigonous, ca 1.5 mm long, apiculate, subsessile.

Ecology: Common in pools, rice-fields, etc.
Fl. & Fr.: Sept. - Nov.
Distribution: Almost throughout India. Sri Lanka, Cochin, China, E. China and Formosa, southwards to Queensland and Malaysia.
Specimens examined: Kabilas, 2009; Silodhar, 2018.

13. Cyperus pygmaeus Rottb. Descr. Icon. Rar. Pl. 20. t. 14. f. 4-5. 1773. *Juncellus pygmaeus* C.B. Clarke in Hook f., Fl. Brit. India 6: 596. 1893; Haines, Bot. Bihar & Orissa pt. 5: 906. 1924; Singh et al., Fl. Bihar 567. 2001. *Cyperus michelianus* var. *pygmaeus* Ascher. & Graebn., Syn. Mitteleur. Fl. 2. 2: 273. 1904; Hooper in Saldanha & Nicolson, Fl. Hassan Dist. 667. 1976.

Annuals, 3-20 cm high. Leaves numerous, flaccid, 1.5-2 mm broad. Spikelets numerous, congested in a terminal, 6-15 mm broad head, ovate, ca 5 x 2.5 mm, 8-20 flowered; bracts 2-6, foliaceous, up to 15 cm long; rachilla sometimes curved or twisted. Glumes with brown streaks, oblong-lanceolate, 2.5-3 x 1 mm, cuspidate, broadly hyaline on margins, stamens usually solitary, rarely 2. Stigmas 2. Nuts orange-brown, ellipsoid, plano-convex, ca 1 mm long (Plate-XXVII; Fig.-30A and B).

Ecology: Along streams and roadside ditches.
Fl. & Fr.: Aug.-Nov.
Distribution: Throughout India.
Specimens examined: Silodhar, 1002; Dhoria, 2022.

14. Cyperus rotundus L., Sp. Pl. 45. 1753; Clarke in Hook, f., Fl. Brit. India 6: 614. 1893; Haines, Bot. Bihar & Orissa pt. 5: 903. 1924. Prain, Bengal Pl. 2: 1145. 1903 (Rep. ed. 2: 862. 1963); Blatter and McCann in J. Bombay Nat. Hist. Soc. 37: 273. 1935; Kuekenthal in Engler, Pfl.-reich 101: 107. 1936; Koyama in Quart. J. Taiwan Mus. 14: 170. 1961 & in Gardens' Bull. Singapore 30: 132. 1977; Kern in Reinwardtia 6:53. 1961 & in van Steenis., Fl. Malesiana ser. 1. 7: 604. 1974; Singh et al., Fl. Bihar 567. 2001; Paria & Chattopadhyay, Fl. Hazaribagh District 2: 996. 2005. *Cyperus rotundus* subsp. *tuberosus* (Rottb.) Kuekenthal in Engler, Pfl.-reich 101: 113. 1936; Kern in Reinwardtia 6: 53. 1961. *Cyperus tuberosus* Rottb. Descr. et. Ic. 28.t. 7. f. 1. 1773; Clarke in Hook. f., Fl. Brit. India 6: 616. 1893; Prain, Bengal Pl. 2: 1145. 1903 (Rep. ed. 2: 862. 1963); Koyama in Quart. J. Taiwan Mus. 14: 170. 1961.

Local name: Motha

Perennials, 10-50 cm high. Stolons long, slender, with intermittent ellipsoid aromatic tubers. Stems compressed-trigonous, sometimes tortuous, base tuberous. Leaves 2-5 mm broad. Anthela usually simple, sometimes compound; bracts 2-4, foliaceous, up to 15 cm long. Spikelets 3-10 together, linear-lanceolate, 10-35 x 1.5-2 mm, frequently curved; rachilla wings with hyaline or red streaks, lanceolate. Glumes tightly imbricating, reddish or deep brown, broadly ovate, 3-3.5 mm long, muticous or mucronulate, keeled. Stamens 3. Styles long exserted; stigmas 3. Nuts glossy black, broadly obovate or obovoid-ellipsoid, trigonous, 1.5-1.7 mm long (Plate-XXVIII; Fig.-31A and 31B).

1 a. Spikelets 1.5-2 mm broad; glumes reddish or deep brown, broadly ovate; nuts obovoid-ellipsoid ...... subsp. *rotundus*

1b. Spikelets 2-2.5 mm broad; glumes pale ferrugineous, deep yellow or reddish yellow, ovate-lanceolate; nuts broadly obovate ..... subsp. *tuberosus*

subsp. **rotundus**

Ecology*:* In open moist place, very common in lawns, etc.
Fl. & Fr.: July - Dec.
Distribution: Throughout India. Cosmopolitan, distributed in tropical, subtropical and temperate regions of the world.
Specimens examined: Asnachuan, 1045; Bukar, 2023.
Uses: Dried roots are used in perfumes; also used in stomach and bowel complaints.

subsp. **tuberosus** (Rottb.) Kuekenth. in Eng., Pflanzenr. heft 101: 113. 1935. *C. tuberosus* Rottb., Icon. Rar. Pl. 28. t.7.f. 1.1773; C.B. Clarke in Hook. f. Fl. Brit. India 6:616. 1893. p.p.

Ecology: Besides rivers and streams.
Fl. & Fr.: Aug.-Jan.
Specimens examined: Danua, 1003; Chordaha, 1070; Garmorwa, 1021.

15. Cyperus tenuispica Steud. Syn. Pl. Cyp. 2: 11. 1854; Singh et al., Fl. Bihar 568. 2001; Paria & Chattopadhyay, Fl. Hazaribagh District 2: 997. 2005. *C. flavidus* C.B. Clarke in J. Linn. Soc. London (Bot.) 21: 122. t. 3. f. 25. 1884 & in Hook. f., Fl. Brit India 6: 600. 1893, non Retz., 1789; Haines, Bot. Bihar & Orissa pt. 5: 893. 1924.

Annual tufted sedge. Stems slender, obtusely trigonous, 10-20 cm high. Leaves shorter or overtopping the stem, 3 mm broad, linear, acute, Inflorescence compound, yellow, finally blackening; brackets 2-3, linear, acuminate, the longest often 16 cm long. Spikelets 2.5-4 mm long, very numerous, linear; rachilla very slender, often reddish. Glumes 0.75 mm long, broadly oblong. Stamens 1, rarely 2; anther linear, muticous, Ovary globose; style and stigmas 0.5 mm long. Nut less than 0.25 mm, globose, obscurely trigonous (Illus. Plate-IV; Fig.-6).

Ecology: Frequent in rice fields and at the edges of tanks.
Fl. & Fr. May-Dec.
Distribution: Throughout India. Sri Lanka, Nepal, from India throughout Indo-China to Malaysia and S. Japan, tropical Australia and tropical Africa.
Specimens examined: Pathalgarwa, 1002; Khairtanr, 2024.

4. FIMBRISTYLIS Vahl, nom. cons.

Key to the species

1a. Nuts linear with almost parallel sides:
Leaves absent; spikelet solitary,
terminal; glumes obtuse ...... 6. *F. tetragona*

1b. Nuts obovoid or suborbicular, with not

parallel sides:

2a. Stigmas 2; nuts lenticular:

3a. Stem bearing 1-3 spikelets:

4a. Glumes ca 1 mm long; stamen 1 ...... 2. *F. argentea*

4b. Glumes 2.5-3 mm long; stamens 1-3: Glumes narrowly oblong, obtuse, twice or more as long as broad ...... 5. *F. polytrichoides*

3b. Stem bearing 4-numerous spikelets: Glumes ca 2 mm long, 3-nerved; nuts without thickened edges ...... 1. *F. alboviridis*

2b. Stigmas 3; nuts trigonal:

5a. Leaves equitant, laterally folded and compressed, the midrib thus appears to be on margins ...... 3. *F. littoralis*

5b. Leaves not equitant, not folded, dorso-ventrally flattened ...... 4. *F. miliacea*

1. Fimbristylis alboviridis C.B. Clarke in Hook. f., Fl. Brit. India 6: 638. 1893; Singh et al., Fl. Madhya Pradesh 3: 296. 2001.

Annuals, 15-30 cm high. Stems compressed-trigonous. Leaves several, 1-2 mm broad; ligule a fringe of hairs. Anthela compound or decompound, lax; bracts foliaceous, spreading. Spikelets solitary, ovoid, terete, 4-11 x 2-2.5 mm. Glumes grayish green or with brown tinge, broadly ovate, 2-2.5 mm long, muticous, 3-nerved. Stamens 1-2. Style 1-1.5 mm long; stigmas 2. Nuts stramineous, obovoid, biconvex, 1-1.2 x 0.8 mm shortly stipitate, verruculose, trabeculate due to the surface cells being arranged in 12-16 rows on each face.

Ecology: Common in open damp places, swamps, wet rice-fields, etc.
Fl. & Fr.: Sept. to Nov.
Distribution: India. Tropical and sub tropical countries.
Specimens examined: Chordaha, 2030; Danua, 1048.

2. Fimbristylis argentea (Rottb.) Vahl, Enum. Pl. 2: 294. 1806; Clarke in Hook. f., Fl. Brit. India 6: 604. 1893; Prain, Bengal Pl. 2: 1154. 1903; Haines,

Bot. Bihar & Orissa pt. 5: 922. 1924; Singh et al, Fl. Bihar 570. 2001; Paria & Chattopadhyay, Fl. Hazaribagh District 2: 1001. 2005. *Scirpus argentea* Rottb., Progr. 1772.

Glabrous annual. Stems densely tufted, setaceous, trigonous, slightly compressed, 3-10 cm long and 0.5-0.7 mm thick. Leaves flat or canaliculated, acuminate, smooth or scabrid at top, shorter than stem, 0.5-0.7 mm wide, with stramineous sheaths; ligule absent. Inflorescence capitate, nearly globose, 0.5-1 cm across, with 4-10 spikelets. Involucral bracts 2-4, dilated at base, the lowest much longer than inflorescence. Spikelets sessile, oblong ovoid or cylindrical, slightly angular, densely many-flowered, 4-10 mm long and 1.2-1.5 mm wide; rachilla narrowly winged. Glumes spiral, broadly ovate-deltoid, with 3-nerved keel, strong midnerve and nerveless, silvery grey or ferrugineous sides, ca 1 mm long and broad. Stamen 1. Style flat, dilated at base, minutely ciliolate above; stigmas 2, shorter than style. Nut broadly obovate or suborbicular, biconvex, shortly stipitate, minutely umbonulate, obscurely reticulate, 0.4-0.5 mm long and broad.

Ecology: Common in swampy places, wet sandy ground of grasslands, rice-fields, etc.
Fl. & Fr.: Nov.-Jan.
Distribution: India: West Bengal, Bihar, Jharkhand, Uttar Pradesh, Central and western India and throughout Deccan Peninsula. Indo-China, Sri Lanka, Thailand and Malaysia.
Specimens examined: Mohane tand, 2067; Chordaha, 1063.

3. Fimbristylis littoralis Gaud., Voy. Uranie 413. 1826; Verma & Chandra in Rec. Bot. Surv. India 21 (2) 256. 1981; Singh et al., Fl. Bihar 572. 2001. *F. miliacea* Vahl., Enum. Pl. 2: 287. 1805 excl. basion; C.B. Clarke in Hook. f., Fl. Brit. India. 6: 644. 1893.

Annuals or perennials, 10-60 (-100) cm high. Stems 4-5 angled. Leaves several, equitant, more or less distichous, frequently yellowish green, laterally flattened, as long as the stem, 1-3 mm broad, finely several striate, thin on outer margin, thick on inner margin, grooved; upper leaves reduced to bladeless sheaths. Anthela compound to supradecompound; bracts short, foliaceous. Spikelets solitrary, sub-globose, 2-5 x 1.5-2 mm, obtuse. Glumes spiral, ovate-oblong, 1-1.5 mm long, muticous, 3-nerved. Stamens 1-2. Styles sparsely fimbriate; stigmas 3. Nuts stramineous, narrowly obovoid, trigonous, 0.6-0.7 x 0.3-0.4 mm, shortly stipitate, verruculose, transversely lineolate.

Ecology: Along ponds and streams.
Fl. & Fr.: Sept.-April.
Distribution: Throughout India.
Specimens examined: Hathia baba, 1118; kabilas, 2070.

4. Fimbristylis miliacea (L.) Vahl, Enum. Pl. 2: 287. 1806; Clarke in Hook. f., Fl. Brit. India 6: 644. 1893; Prain, Bengal Pl. 2: 1155. 1903(Rep. ed. 2: 869. 1963); Haines. Bot. Bihar & Orissa pt. 5. 915. 1924; Blatter and McCann in J. Bombay Nat. Hist. soc. 37: 546. 1935; Kern in Reinwardtia 6: 40. 1961; Koyama in J. Fac. Sci. Univ. Tokyo sect. 3. 8: 108. 1961 & in Bot. Mag. Tokyo 87: 316. 1974; Singh et al., Fl. Bihar 572. 2001; Paria & Chattopadhyay, Fl. Hazaribagh District 2: 1006. 2005. *Scirpus miliaceus* L., Syst. Veg. ed. 10. 686. 1759.

Glabrous annual. Stems erect, densely tufted acutely 4-5 angled, striate, 10-50 cm tall and 1-2 mm thick; the base clothed with obliquely truncate bladeless sheaths. Leaves dorsiventrally flattened, acuminate, scabrid on rib-like angles in upper part, shorter than to as long as stems, 2-3 mm wide; ligule absent. Inflorescence compound or decompound, loose, 4-10 cm long. Involucral bracts 3-5, setaceous, to 3 cm; primary rays several, compressed, scabrid, to 6 cm. Spikelets solitary, ovoid-globular, rusty brown, 2-4 mm, densely many-flowered; rachilla narrowly winged. Glumes spiral, broadly ovate, mucronate, with 3-nerved keel and hyaline margins, 1-1.3 mm long and 0.8-1 mm wide. Stamen usually 1. Style triquetrous, pyramidally thickened at base; stigmas 3, as long as style. Nut broadly obovoid, trigonous, verruculose, transversely lineolate, stramineous, 0.5-0.7 mm long and 0.4-0.5 mm wide.

Ecology: Very common; in open wet places, rice-fields, swampy grassland, etc.
Fl. & Fr.: Aug.-Dec.
Distribution: Throughout India. Tropical and subtropical regions of whole world; in eastern Asia the range extends into temperate regions as far as to central Japan and China.
Specimens examined: Duragara, 1018; Mainukhar, 2074.

5. Fimbristylis polytrichoides (Retz.) R. Br., Prodr. 226. 1810; C.B. Clarke in Hook. f., Fl. Brit. India 6: 623. 1893. *Scirpus polytrichoides* Retz., Observ. Bot. 4: 11. 1786.

Annuals or perennials, slender, 5-30 cm high. Leaves about half as long as the stem; sheaths glabrous or puberulous. Spikelets solitary (rarely 2-3), terminal, oblong-ellipsoid, 5-8 mm long; foliaceous bracts absent but sometimes the lowest empty glume with a leaf-like, up to 25 mm long appendage resembling a bract or an extension of the stem. Glumes pale brown or whitish, sometimes reddish at apex, narrowly oblong, nearly flat in flowers, navicular in fruits, ca 2.5 mm long, slightly keeled, obtuse. Stamens 1-3. Stigmas 2. Nuts pale or dark brown obpyriform, biconvex, 1-1.5 mm long, subsessile, minutely papillose; papillae often whitish.

Ecology: Common in grassy rocky fields.
Fl. & Fr.: July-Sep.
Distribution: Throughout India.
Specimens examined: Chamargadda, 1064; Garmorwa, 2073.

6. Fimbristylis tetragona R. Br., Prodr. Fl. Nov. Holl. 226. 1810; Clarke in Hook. f., Fl. Brit. India 6: 631: 1893; Prain, Bengal Pl. 2: 1152. 1903; Haines, Bot. Bihar & Orissa pt. 5: 918. 1924; Singh et al., Fl. Bihar 574. 2001; Paria & Chattopadhyay, Fl. Hazaribagh District 2: 1007. 2005.

Annuals or perennials, 10-60 cm high. Stems tetragonous. Leaves absent; sheaths 2-3, pale or chestnut brown (sometimes the uppermost with a short blade). Spikelets solitary, terminal, erect, ovoid-globose, terete, 6-18 x 4-6 mm, acute. Glumes spiral, stramineous or with brown tinge, ovate-oblong, 3-5 mm long, obtuse. Stamens 1-3. Stigmas 2-3. Nuts stramineous, linear-oblong with almost parallel sides, 1.5-2 x 0.5 mm long, stipitate, trabeculate due to the surface cells being arranged in about 9 vertical rows.

Ecology: Common in swamps, wet rice-fields, etc.
Fl. & Fr.: Aug.-Nov.
Distribution: Throughout India except north-west regions. Tropical southern Asia from India and Sri Lanka through Indo-China northeastwards to southern China, eastwards to Malaysia and northern Australia.
Specimens examined: Mohane, 2089; Garmorwa, 2051.

5. **FUIRENA** Rottb.

Fuirena ciliaris (L.) Roxb., Fl. Ind. 1: 184. 1820; Kern in van Steenis, Fl. Males ser. 1. 7: 519, f. 32. 1974; *Scirpus ciliaris* L., Mant. 2: 182. 1771; Singh et al., Fl. Bihar 575. 2001; Paria & Chattopadhyay, Fl. Hazaribagh District 2: 1008. 2005. *Fuirena glomerata* Lam., Tabl. Enc. Meth. Bot. 1: 150. 1791; Clarke in Hook. f., Fl. Brit. India 6: 666. 1893; Prain, Bengal Pl. 2: 1158. 1903; Haines, Bot. Bihar & Orissa pt. 5: 928. 1924. *F. rottboelli* Nees in Wight, Contrib. Bot. India 94. 1834.

Hairy annual; stem 14.2-58.8 cm high, striate. Leaves grass like, 5.2-14.2 x 0.3-0.8 cm, acuminate, margins ciliolate; sheaths long, closed, striate. Spikelets 1-8, 7.5 mm long, ovoid to oblong obtuse, greenish-brown; bracts 0. Glumes 1.5-1.8 mm long (excluding the arista), imbricate, sparsely hairy without obovate at length, deciduous; keel ending in 0.8-1.1 mm long hairy apical arista. Hypogynous scales 0.8-1.2 mm long, membranous, persistent, quadrate with a slender claw; apex minutely acutely 3-toothed with the middle tooth slightly longer. Nut 0.7-0.9 mm long, elliptic-obovoid, light brown, minutely apiculate.

Ecology: Common in rice-fields, margins of ponds, ditches, etc.
Fl. & Fr.: October - January.
Distribution: Throughout India and Old World Tropics.
Specimens examined: Duragara, 2032; Pathalgara, 2073.

6. **KYLLINGA** Rottb., nom. cons.

Key to the Species

1a. Keel of the glumes broadly winged ...... 3. *K. nemoralis*
1b. Keel of the glumes not winged:
  2a. Rhizome short stolons wanting ...... 2. *K. bulbosa*
  2b. Rhizome long, horizontally creeping ...... 1. *K. brevifolia*

1. Kyllinga brevifolia Rottb., Descr. & Ic. Rar. Nov. Pl. 13, t. 4. f. 3. 1773; Clarke in Hook. f., Fl. Brit. India 6: 588. 1893; Prain, Bengal Pl. 2: 1135; Haines, Bot. Bihar & Orissa pt. 5: 897. 1924; Singh et al., Fl. Bihar 576.

2001; Paria & Chattopadhyay, Fl. Hazaribagh District 2: 1009. 2005. *Cyperus brevifolius* (Rottb.) Hassk., Cat. Hort. Bog. 24. 1844; Rao & Verma, Cyper. N.E. India 6. 1982.

Perennials. Rhizome 3.8-18.2 cm long, stoloniferous, creeping; stems many, usually at distant, 10-34 cm high, compressed-trigonous. Leaves usually distinctly shorter than the stem, 1.5-3.2 mm broad, acuminate. Head solitary, up to 6 mm long, green, globose; bracts 3-4 leaf-like. Spikelets up to 3 mm long, elliptic-oblong to oblong-lanceolate, compressed, 1-flowered. Glumes: 2, lower one minute, empty; 3rd one flower-bearing, ovate-lanceolate with a green mucronate keel, usually 3-nerved on the wings, scarious margined, 4th glume empty. Nut 1.1-1.3 mm long, elliptic-obovate, slightly compressed, light yellowish or brownish-yellow.

Ecology: Common along the paddy fields and the sides of water courses.
Fl. & Fr.: Throughout year especially from June to November.
Distribution: Throughout India; common in all warm parts except Mediteranean.
Specimens examined: Bukar, 2084; Sikda, 2037.

2. Kyllinga bulbosa Beauv., Fl. d'Oware & Benin 1: 11, t. 8. f. 1. 1804; Koyama in Dassan. & Fosb., Rev. Handb. Fl. Ceylon 5: 245. 1985; Paria & Chattopadhyay, Fl. Hazaribagh District 2: 1010. 2005. *Schaenoides triceps* Rottb., Descr. Pl. Rar. Progr. 15. 1772, nom. inval. *Kyllinga triceps* Rottb., Descr. & Ic. Rar. Nov. Pl. 14. t. 4. f. 6. 1773, nom. illegit.; Clarke in Hook. f., Fl., Brit. India 6: 587. 1893; Prain, Bengal Pl. 2: 1135. 1903; Haines, Bot. Bihar & Orissa pt. 5: 907. 1924.

Local name: Nirbisi.

Perennials. Rhizome short; stolons absent. Stems densely tufted, slender, sometimes setaceous, obtusely trigonous, smooth, 5-20 cm tall and 0.6-1 cm wide, with increassate base covered by brownish sheath. Leaves flat or slightly conduplicate, gradually acuminate, nearly as long as stem. 1.5-2.5 mm wide. Inflorescence capitate, consisting of usually 3 (rarely 1, 4 or 5) sessile, dense spikes; central spike subglobose, 5-8 cm long and 4-6 cm wide; lateral ones globose, smaller. Involucral bracts 3 or 4, spreading or reflexed, to 10 cm long. Spikelets oblong, compressed, pale-green, 1.6-2 mm long and 0.6-0.7 mm wide. Glumes ovate-oblong, hyaline with sharp, almost smooth keel; 1st and 2nd glume small. 3rd 7-nerved, 1.5-1.7 mm long, 4th 5-nerved, 1.7-2 mm long. Stamens 2. Nut biconvex, laterally compressed, oblong, brownish, 1-1.2 mm long and 0.5mm wide.

Ecology: Common in waste places, road-sides, open grasslands, cultivated fields, forests etc.
Fl. & Fr.: Aug.-Feb.
Distribution: Throughout India. From tropical Africa through Pakistan, India and Indo-China northeastwards to southern China and eastwards to Malaysia and northern Australia.
Specimens examined: Sanjha, 2072; Siarkoni, 2041.
Uses: An oil extracted from roots is used for stimulating liver and to relieve pruritus.

3. Kyllinga nemoralis (J. R. & G. Forster) Dandy ex Hutchinson & Dalziel, Fl. W. Trop. Africa ed. 1. 2: 487. 1936; Koyama in Gardens' Bull. Singapore 30: 163. 1977; Singh et al., Fl. Bihar 576. 2001; Paria & Chattopadhyay, Fl. Hazaribagh District 2: 1012. 2005. *Thyrocephalon nemorale* J. R. & G. Forster, Char. Gen. Pl. 130. 1776. *Kyllinga monocephala* Rottb., Descr. et Icon, 13. t. 4. f. 4. 1773: Partly, excl. syn., nom. illeg.; Clarke in Hook. f., Fl. Brit. India 6: 588. 1893; Prain, Bengal Pl. 2: 1135. 1903(Rep. ed. 2: 855. 1963) (*microcephala*); Haines, Bot. Bihar & Orissa pt. 5: 907, 1924; Blatter and McCann in J. Bombay Nat. Hist. Soc. 37: 25. 1935. *Cyperus kyllinga* Endl., Cat. Hort. Acad. Vindb. 1: 94. 1842; Kuekenthal in Engler. Pfl.- reich 101: 606. 1936; Koyama in Quat. J. Taiwan Mus. 14: 192. 1968; Kern in Reinwardtia 6: 67. 1961.

Perennials. Rhizome short; stem 7-36 cm high, usually solitary, erect, compressed-triangular. Leaves usually as long as or in many cases longer than the stem, 2-4 mm broad, acuminate at apex, mid nerve prominent. Head up to 7 mm long, spiciform, globose to broadly-oblong, more or less white; bracts 3-4, leaf-like, longer one up to 20 cm. Spikelets many, up to 2 mm long, 1 or rarely 2-flowered, bisexual or sometimes male by abortion. Glumes: outer 2 hyaline, lowest 1.3-1.8 mm long, lanceolate, second broadly ovate, obtuse; 3rd and 4th almost green, falcately incurved, acuminate, upper slightly longer, 3rd broader, keel dorsally spinulosely distinctly or sometimes obscurely winged. Nut 0.8-1.2 mm long, obovoid-oblong, minutely punctate, apiculate.

Ecology: Common, preferably in dampy places.
Fl. & Fr.: July-Feb.
Distribution: India and Sri Lanka. Common in hot and warmer regions of the Old World.
Specimens examined: Muria, 2066; Dhoria, 2053.

7. MARISCUS Vahl

Key to the Species

1a. Anthelas compound or decompound; Spikes subglobose: Spikelets 6-12 mm long, each maturing 4-14 nuts; glumes muticous ...... 1. *M. compactus*
1b. Anthelas simple; spikes cylindrical: Rhizome emitting slender stolons; Spikelets bearing 1-4 nuts ...... 2. *M. paniceus*

1. Mariscus compactus (Retz.) Boldingh, Zakfl. Landb. Java 77. 1916; Haines, Bot. Bihar & Orissa 5: 910. 1924 (Repr. ed., 3: 953. 1961); Singh et al., Fl. Bihar 578. 2001. *Cyperus compactus* Retz., observ. Bot. 5: 10. 1789. *Mariscus microcephalus* Presl. Reliq. Haenk, 1: 182. 1830; C. B. Clarke in Hook. f., Brit. India 6: 624. 1893.

Perennials, 0.2-1 m high. Stems spongy, subterete or trigonous, 2-6 mm thick. Leaves 3-11 mm broad, with transverse nervules more or less distinctly raised, compound or decompound, young greenish, mature bright reddish brown, stellately sprading, linear-lanceolate, 6-12 x 1-1.5 mm rachilla hyaline-winged. Glumes oblong-lanceolate, 3-4 mm long, obtuse; keel green; sides reddish brown on maturity. Stamens 3. Styles 3-fid. Nuts deep brown, linear-oblong, trigonous, slightly curved, 1.5-2 mm long, apiculate or rostrate, densely puncticulate.

Ecology: Along streams, rivers, etc.
Fl. & Fr.: Sept.-Jan.
Distribution: Almost throughout India.
Specimens examined: Chordaha, 1092; Danua, 1046.

2. Mariscus paniceus (Rottb.) Vahl, Enum. Pl. 2: 378. 1806; C. B. Clarke in Hook. f., Fl. Brit. India 6: 620. 1893; Haines, Bot. Bihar & Orissa pt. 5: 1257. 1924 (Repr. Ed., 3: 1311. 1961); Singh et al., Fl. Bihar 579. 2001. *Kyllinga panicea* Rottb. Descr. Icon. Rar. Pl. 15. t. 4. f. 1. 1773.

Perennials. Stolons subterranean, scaly. Stems 10-50 cm high, trigonous. Leaves shorter or longer than stem, linear, acuminate, 3-4.5 mm broad, 1-nerved. Inflorescence a simple umbel of spikes 0.5-1.5 cm long. Spikelets 1-flowered; bracts 3-6, linear, 5-20 cm long; rays 3-7, 0.2-3.5 cm long. Glumes apparently 4; 2 lowest ones empty; 3rd glume ovate, fertile, 3-3.5 mm long, obtuse. Stamens 3. Nut oblong, pale brown, trigonous (Plate-XXVIII; Fig.-32).

Ecology: Occasionally found in wet places.
Fl. & Fr.: July-Sept.
Distribution: Almost throughout India.
Specimens examined: Sanjha, 1004; Siarkoni, 1060; Mainukhar, 1062.

8. PYCREUS Beauv.

Key to the Species

1a. Midvein of the glumes excurrent beyond glume apex into a distinct mucro ...... 2. *P. pumilus*

1b. Glumes muticous: Stem erect, not rooting at nodes, leafy only at very base. Glumes with straight keel and without depressions ...... 1. *P. flavidus*

1. Pycreus flavidus (Retz.) Koyama in J. Jap. Bot. 51: 313. 1976 & in Dassan. & Fosb., Rev. Handb. Fl. Ceylon 5: 222. 1985; Singh et al., Fl. Bihar 580. 2001; Paria & Chattopadhyay, Fl. Hazaribagh District 2: 1016. 2005. *Cyperus flavidus* Retz., Obs. Bot. 5: 13. 1788; Rao & Verma, Cyper. N. E. India 10.1982. *Cyperus globossus* All, Fl. Pedem. Auct. 49. 1789, non Forsk. 1775. *C. capillarris* Koen, ex Roxb., Fl. Ind. 1: 198. 1820. *Pycreus globossus* (All.) Reichb., Fl. Germ. Excurs. 2: 130. 1830; Haines, Bot. Bihar & Orissa pt. 5: 905. 1924.

P. capillaris (Koen. ex Roxb.) Nees ex Clarke in Hook. f. Fl. Brit. India 6: 519. 1893, incl. vars.; Prain, Bengal Pl. 2: 1137. 1903.

Annual, or perennial with short rhizome, Stems erect, slender, tufted, trigonous, leafy only at very base, 5-60 cm tall and 1-1.5 mm thick, Leaves narrow, canaliculated, setaceous, smooth or minutely scabrid at top, shorter than stems, 1-2.5 mm wide; lower sheaths ferruginous to reddish-brown. Inflorescence simple or sub compound, open to contracted into a single cluster. Involucral bracts 2-4, lowest 1-2 much overtopping the inflorescence to 25 cm long. Primary rays 3-6 cm long. Spikes ovoid with 5-20 spikelets. Spikelets linear, parallel-sided, compressed, 1-2 cm long and 2.5-3 mm wide, 20-40 flowered; rachilla straight, wingless, persistent. Glumes oblong-ovate, 1.5-2.5 mm long and 1-1.5 mm wide, with green, straight, 3-nerved keel, nerveless stramineous brown sides and whitish hyaline margins. Stamens 2. Style short; stigmas 2, longer than style. Nut obovate to oblong-elliptic, biconvex, laterally compressed, apiculate, dark brown when mature, 0.8-1.2 mm long and 0.4-0.6 mm wide.

Ecology: Common in open, wet places, grasslands, rice-fields, etc.
Fl. & Fr.: Feb.-May
Distribution: Throughout India. Widely distributed from the mediterra-nean region and tropical Africa eastwards to Central and South Asia to Malaysia and Australia and northwards to Japan.
Specimens examined: Khairtanr, 1120; Sanjha, 1088.

2. Pycreus pumilus (L.) Nees ex Clarke in Hook. f. Fl. Brit. India 6: 591. 1983, quoad nom. cit. excl. basion; Haines, Bot. Bihar & Orissa pt. 5: 905. 1924; Koyama in Dassan. & Fosb., Rev. handb. Fl. Ceylon 5: 224. 1985; Singh et al., Fl. Bihar 580. 2001; Paria & Chattopadhyay, Fl. Hazaribagh District 2: 1017. 2005. *Cyperus pumilus* L., Cent. Pl. 2: 6. 1756; Kern in Steenis, Fl. Males. ser. 1, 7: 650. 1974. *Cyperus nitens* Retz., Obs. Bot. 6: 13. 1789. *Pycreus nitens* (Retz.) Nees in Nova Act. Acad. Caes. Leop.- Carol. Nat. Cur. 19, suppl. 1: 53. 1843; Clarke in Hook. f., Fl. Brit. India 6: 591. 1863; Prain, Bengal. Pl. 2: 1137. 1903.

Caespitose annual; stem up to 20 cm long compressed-trigonous. Leaves 2.8-12.2 x 0.1-0.14 cm. Spikelets clustered into simple heads or in condensed spikes in few rays and central sessile clusters surrounded with 3.5 up to 15 cm long leaf like bracts, 3.8-8.8 x 2.1-2.4 mm, compressed, oblong or linear-oblong, obtuse or sub-acute, 10-24 flowered, rachilla exaltate. Glumes densely imbricate, 1.3-1.8 mm long, navicular, margin hyaline, mucro conspicuous, slightly recurved. Nut 0.4-0.6 mm long, light brown, punctulate, broadly obovoid, slightly compressed, minutely apiculate.

Ecology: very common in damp localities, rice-fields, etc.
Fl. & Fr.: July-Nov.
Distribution: Throughout India; warmer regions of the world.
Specimens examined: Muria, 2052; Silodhar, 2071.

9. SCHOENOPLECTUS (Reichb.) Palla, nom. cons.

Key to the Species

1a. Stems and bracts transversely septate; the bract longer than or about as long as stem proper, hence inflorescence seemingly inserted in lower part or about middle of stem: Glumes not inflated in fruit, not shining. Bract longer than stem proper ...... 1. *S. articulatus*

1b. Stems and bracts not septate; the bract shorter than stem proper, hence inflorescence inserted in upper part of stem ...... 2. *S. supinus*

1. Schoenoplectus articulatus (L.) Palla in Bot. Jahrb. Syst. 10: 229. 1889; Koyama in Dassan. & Fosb., Rev. Handb. Fl. Ceylon 5: 163. 1982; Singh et al., Fl. Bihar 582. 2001; Paria & Chattopadhyay, Fl. Hazaribagh District 2: 1021. 2005. *Scripus articulatus* L., Sp. Pl. 47. 1753; Clarke in Hook. f., Fl. Brit. India 6: 656. 1893; Prain, Bengal Pl. 2: 1160. 1903; Haines, Bot. Bihar & Orissa pt. 5: 926. 1924; Rao & Verma, Cyper. N.E. India 43. 1982.

Local name: Chichora.

Perennials. Stems erect, tufted, terete, fistulose, transversely septate, smooth, 10-50 cm tall and 3-8 cm wide, clothed at base with 2 or 3 bladeless sheaths only. Upper 2 sheaths cylindrical, stramineous-brown, scarious-margined, obliquely truncate, 4-15 cm long; lower 1 or 2 reduced and scale-like, 1-1.2 cm long, brown. Inflorescence pseudolateral, capitate, globose, 1-3 cm across, consisting of numerous (up to 50 cm) spikelets. Involucral bract 1, similar to and continuous with stem, terete, transversely septate, somewhat to much longer than stem proper, to 60 cm. Spikelets sessile, ovoid to oblong-ovoid, terete, acutish, 0.7-1.5 cm long and 3-5 mm wide, densely many-flowered. Glumes appressed, concave, rusty or purple, ovate-suborbicular, apiculate, hardly keeled, many-nerved, 3-5 mm long and wide. Hypogynous bristles 0. Stamens 3. Stigmas 3, shorter than style. Nut triquetrous, obovoid, apiculate, transversely wavy-ridged, black, 1.2-1.5 mm long and 0.8-1 mm wide (Illus. Plate-V; Fig.-7).

Ecology: Very common; in open swampy or inundated places, shallow pools, rice-fields, margins of ponds, etc.
Fl. & Fr.: Nov.-Jan.
Distribution: Throughout India. Indo-China, Malaysia and tropical Australia.
Specimens examined: Mohane tand, 1038, Muria, 2044.
Uses: Dried plants are used for thatching; Rhizomes are given in diarrhea and to stop vomiting.

2. Schoenoplectus supinus (L.) Palla ssp. *lateriflorus* (Gmel.) Koyama in Hara et al., Enum. Fl. Pl. Nepal 1: 119. 1978 & in Dassan. & Fosb., Rev. Handb. Fl. Ceylon 5: 158. 1985; Singh et al., Fl. Bihar 584. 2001; Paria & Chattopadhyay, Fl. Hazaribagh District 2: 1023. 2005. *Scirpus lateriflorus* Gmel., Syst. Veg. 127. 1791. *S. supinus* auct. non L.: Clarke in Hook f., Fl. Brit. India 6: 655. 1893; Prain, Bengal. Pl. 2: 1160. 1903; Haines, Bot., Bihar & Orissa pt. 5: 925. 1924. *S. supinus* L. var. *uninodes* sensu Clarke in Hook. f., Fl. Brit. India 6: 656. 1893, quaod nom. cit. non *Isolepis uninoides* Delile; Prain, Bengal Pl. 2: 1160. 1903.

Annuals; stems tufted, 7.5-30.5 cm high, slender, faintly trigonous, striate. Leaves very short or nil; sheaths short or long, mouth oblique, rarely produced into a short acute limb. Spikelets 2-7, in a cluster coming out laterally about the middle or little above the middle of the stem, up to 5.2 mm long, ovate, ellipsoid or oblong, sessile or sub-sessile; rachilla slender, not winged. Glumes 1.8-2.6 mm long, closely imbricate, membranous, ovate, acute, keel prominent, produced into a short mucro. Hypogynous bristles nil. Nut 0.8-1.3 mm long, transversely lineolate broadly obovoid, apiculate, trigonous, one face broader.

Ecology: Common in ponds, ditches, etc.
Fl. & Fr.: September-February.
Distribution: India, Sri Lanka, Tropical Asia, N. America and Australia.
Specimens examined: Murtiakalan, 3001; Silodhar, 2054.

2. POACEAE

Key to the Genera

1a. Plants tall, perennial; culms woody; leaf blades flat, lanceolate to ovate; ligules scarious; lodicules usually 3; stamens 6:

2a. Spikelets not in globose heads; palea keeled, ciliate or scabrid; pericarp thin, adnate to the grain, adhering to hilum only:
Spikelets many-flowered; palea scabrid; filaments free ...... 4. *Bambusa*

2b. Spikelets in globose heads; palea neither keeled nor ciliate; pericarp crustaceous or fleshy, separable from the grain:
Two-three central florets bisexual; lodicules absent; fruits crustaceous ...... 13. *Dendrocalamus*

1b. Plants small, annual or perennial: culm herbaceous; leaf blades flat or filiform, usually linear to lanceolate; ligules ciliate or membranous; lodicules 2 or absent; stamens 1-3 (6 in *Oryza*):

3a. Spikelets unisexual, male and female conspicuously dissimilar:

4a. Female spikelets completely enclosed in a metamorphosed leaf sheath, forming spherical or cylindrical bead like structure ...... 9. *Coix*

4b. Female spikelets not enclosed in a false Involucre as above;
Male and female spikelets in separate inflorescences; female spikelets in longitudinal rows on a very thick axis; male spikelets in a large terminal panicle ...... 49. *Zea*

3b. Spikelets perfect with one floret, If unisexual then male and female spikelets similar, with or without reduced florets, occasionally 2 or more perfect florets:

5a. Spikelets in pairs, one sessile, other pedicelate, rarely solitary and all alike; first glume large and firm, clasping or enclosing the florets:

6a. Joints of the rachis and pedicel narrow, slender, if thickened then spikelets 1-flowered and awned:

7a. Pairs of spikelets alike and with perfect florets (rarely in threes or the pedicillate reduced):

8a. Rachis persistent or tardily breaking up;
Panicle narrow, contracted: glumes very delicate; callus hairs twice or more longer than glumes ...... 25. *Imperata*

8b. Rachis disarticulating: Spikelets awnless; upper lemma not clefted ...... 40. *Saccharum*

7b. Pairs of spikelets dissimilar; sessile bisexual; pedicellate male or sterile:
Racemes paired, digitate, scattered or in panicle:

9a. Racemes espatheate:

10a. Racemes arranged in panicle or in whorls on main axis:

11a. Racemes with many pairs of spikelets ...... 48. *Vetiveria*

11b. Racemes with 1 sessile and 2 pedicellate spikelets ...... 8. *Chrysopogon*

10b. Racemes digitate, subdigitate or scattered:

12a. Sessile spikelets of all pairs bisexual, awned; lower glume of sessile spikelet narrow and not closely imbricating ...... 5. *Bothriochloa*

12b. Sessile spikelets of lowest 1-3 pairs male or neuter, awnless; lower glume of sessile spikelet broad and closely imbricating ...... 15. *Dichanthium*

9b. Racemes spatheate: Plants aromatic; racemes paired in compound panicle; one pair of spikelet in each receme homogamous ...... 10. *Cymbopogon*

8b. Racemes solitary:

13a. Racemes of several pairs of evenly placed spikelets, without an involucres: rhachis tough, column of awn hairy ...... 23. *Heteropogon*

13b. Racemes of a few pairs of spikelets,

the lower pairs male or neuter, forming
a false involucres around others:

14a. Involucral spikelets persistent, shortly pedicellate or sessile; callus of fertile spikelets pointed ...... 45. *Themeda*

14b. Involucral spikelets deciduous, with long pedicels; callus of fertile spikelets truncate or obtuse ...... 27. *Iseilema*

6b. Joints of the rachis and pedicel swollen,
3-angled, rounded or flattened:

15a. Sessile spikelet with a male and bisexual floret;
Upper lemma usually awned:

16a. Racemes many-noded not
enclosed in a spathe like sheath:

17a. Racemes solitary; lower
glume of sessile spikelet
without nodules or furrows ...... 42. *Sehima*

17b. Racemes 2 to many; lower glume of
sessile spikelet with nodules or transverse
furrows; Pedicellate spikelets developed;
lower glume coriaceous below
transversely wrinkled, with nodules
on margins ...... 26. *Ischaemum*

16b. Racemes 1-noded, reduced to
3 heteromorphous spikelets,
enclosed in cymbiform spathe ...... 2. *Apluda*

15b. Sessile spikelet with a bisexual floret

only or occasionally with a male floret below; upper lemma usually awnless:

18a. Sessile spikelets spherical; lower Glume pitted all over; pedicellate Spikelet reduced ...... 22. *Hackelochloa*

18b. Sessile spikelet not spherical; lower glume not pitted; pedicellate spikelet well developed or absent; Spikelets paired, one sessile, the other pedicellate ...... 39. *Rottboellia*

5b. Spikelets solitary or paired, if paired then first glume not larger and firmer than lemma of fertile floret, not clasping or enclosing second glume:

19a. Reduced floret (male or female) always below the perfect one:

20a. Reduced floret 1, its lemma similar to second glume; disarticulation below the glume or glumes:

21a. Spikelets in involucres of bristles or subtended by solitary bristle, and falling with or without bristles at maturity:

22a. Upper lemma smooth; bristles disarticulate with spikelets; Bristles many modified or spine like branchlets; Involucre of free, naked or plumose bristles ...... 35. *Pennisetum*

22b. Upper lemma transversely rugose; bristles persistent ...... 43. *Setaria*

21b. Spikelets not in involucres or subtended by bristles: Spikelets awnless, if awned, then subsessile in false, second, variously arranged spikes and awns from the tips of glume and lemma; Upper lemma crustaceous, rarely chartaceous:

23a. Inflorescence an open or contracted panicle, sometimes spike like panicle:

24a. Spikelets arranged in open or contracted panicle; upper glume not inflated: spikelets not or slightly gibbous, not much compressed ...... 32. *Panicum*

24b. Spikelets arranged in cylindrical spike like panicles; upper glume inflated ...... 41. *Sacciolepis*

23b. Inflorescence one-sided spikes or spike like racemes, digitate or scattered, rarely solitary:

25a. Lemma of the upper floret thinly cartilaginous, usually with flat margins:

26a. Spikelets awnless ...... 16. *Digitaria*

26b. Spikelets awned ...... 1. *Alloteropsis*

25b. Lemma of the upper floret crustaceous or coriaceous, with inrolled margins:

27a. Lower glume and lowest internode of the rachilla forming a swollen callus at the base of the spikelet ...... 21. *Eriochloa*

27b. Lower glume and lowest internode of rachilla not forming a swollen callus at the base of the spikelets:

28a. Spikelets adaxial; lower glume turned towards the rachis, back of upper lemma turned away from it ...... 6. *Brachiaria*

28b. Spikelets abaxial; lower glume (if present) turned away from the rachis, back of upper lemma facing it:

29a. Spikelets plano-convex; lower glume usually absent ...... 34. *Paspalum*

29b. Spikelets otherwise; lower glume developed, sometimes minute:

30a. Glumes acuminate or awned; upper lemma not mucronate:

31a. Leaf blades linear; ligule absent; racemes dense ......17. *Echinochloa*

31b. Leaf blades lanceolate to ovate; ligule present; racemes loose ...... 30. *Oplismenus*

30b. Glumes awnless, if acuminate the upper lemma mucronate: ...... 33. *Paspalidium*

20b. Reduced floret 1-2, its lemma dissimilar to Second glume; disarticulation above the glumes:

32a. Glumes absent or rudimentary:

Plants not floating; leaf blades linear; glumes present:

33a. Annuals; spikelets 7-10 mm long; cultivated ...... 31. *Oryza*

33b. Perennials; spikelets less than 6 mm long; wild ...... 28. *Leersia*

32b. Glumes present, at least the second well developed:

34a. Spikelets 3-flowered, strongly compressed, lower 2 represented by empty lemmas ...... 37. *Phalaris*

34b. Spikelets 1or 2 flowered, not compressed:

35a. Spikelets 2-flowered, lower male or barren, upper bisexual:
Lower floret empty, without palea; rachilla produced beyond the upper floret; glumes up to half the length of spikelet ...... 46. *Thysanolaena*

35b. Spikelets with one perfect floret, usually reduced floret absent:

36a. Spikelets 3-awned; lemmas Indurated, rigid at maturity ...... 3. *Aristida*

36b. Spikelets not as above; lemmas hyaline or membranous:

37a. Glumes awned; lemmas 3-5 nerved; caryopsis with an

adhering pericarp ...... 38. *Polypogon*

37b. Glumes awnless; lemmas 1-3 nerved; caryopsis with free pericarp: Culms erect; inflorescence a spreading panicle, not subtended by inflated sheath ...... 44. *Sporobolus*

19b. Reduced florets absent, if present always above perfect one:

38a. Inflorescence a panicle or open raceme, primary branches spreading or contracted but not spicate:
Plants less than 1.5 m long, not rhizomatous:
Lemmas awnless ...... 20. *Eragrostis*

38b. Inflorescence a spike or spicate raceme, or with 2 to several spicate primary branches:

39a. Inflorescence of 1 to several, unilateral, spicate primary branches:
Spikelets not in deciduous bur like clusters;
glumes without hooked spines:

40a. Spikelets with 2 or more perfect florets:

41a. Inflorescence branches paired, verticillate or clustered at culm apex:

42a. Rachis projecting in a sharp point ...... 12. *Dactyloctenium*

42b. Rachis ending in a spikelet:

43a. Glumes or lemmas awnless, tips not toothed ...... 18. *Eleusine*

43b. Glumes or lemmas awned, if awnless then the tip of lemma toothed: Lemma tip entire, 2-3 awned ...... 7. *Chloris*

41b. Inflorescence branches distributed along the culm axis: Glumes awnless, obtuse or acute:

44a. Culms with extensive rhizomes; emerging leaf piercing as needle; inflorescence a loose or contracted panicle ...... 14. *Desmostachya*

44b. Culms without rhizomes; Emerging leaf soft; inflorescence of spikes or racemes: racemes with many, secund, laterally compressed spikelets; lemma keeled ...... 29. *Leptochloa*

40b. Spikelets with 1 perfect floret:

45a. Inflorescence of solitary raceme; spikelets awned ...... 36. *Perotis*

45b. Inflorescence of 2-6 digitate spikes; spikelets awnless ...... 11. *Cynodon*

39b. Inflorescence a terminal, bilateral spike or spicate raceme:

46a. Spikelets solitary at each node of the inflorescence axis:

47a. Spikelets awnless, arranged in 2 rows on a long, terminal, slender axis ..... 19. *Eragrostiella*

47b. Spikelets awned, not in 2 rows; Fertile lemma longer than glumes, entire; awns more than 3 cm long ...... 47. *Triticum*

46b. Spikelets 2-3 at each node of the inflorescence axis; Spikelets not embedded in the rachis; glumes standing in front of the spikelet ...... 24. *Hordeum*

1. ALLOTEROPSIS C. Presl

Alloteropsis cimicina (L.) Stapf in Prain, Fl. Trop. Afr. 9: 487. 1919; Haines, Bot. Bihar & Orissa pt. 5: 1009. 1924; Bor, Common Grasses of the United Provinces 66. 1940; Bor, Grass. Burm. Ceyl. Ind. Pak. 276. 1960; Sreekumar & Nair, Fl. Kerala Grasses 214. 1991; Singh et al, Fl. Bihar 588. 2001; Paria & Chattopadhyay, Fl. Hazaribagh District 2: 1036. 2005; Kabir & Nair, Fl. Tamil Nadu Grasses 211. 2009. *Milium cimicinum.* L., Mant. Pl. 184. 1771. *Axonopus cimicinus* P. Beauv., Ess. Agrostogr. 12. 1812; Hook. f., Fl. Brit. India 7: 64. 1896.

A tufted annual, 30-60 cm high. Culms erect or decumbent at the base, hairy. Leaves 3.2-7 x 0.9-2.5 cm, ovate-lanceolate, flat, acute, glabrous or hairy, margins ciliate with stiff bulbous-based hairs; ligule shortly hairy. Inflorescence made up of 3-8 spike-like recemes; pedicel 1-3 mm long, cupular at the tip. Spikelets 4 mm long, ovoid, or elliptic, 1-2 nate. Gl. I ovate-lanceolate, acuminate, 2.5 mm long, glabrous, 3-nerved. Gl. II ovate-lanceolate, 4 mm long, 5-nerved, shortly aristate. Lower lemma male, 3.5 mm long, 5-nerved; palea 2-partite, hyaline. Upper lemma bisexual, ovate-oblong, 3 mm long, abruptly awned, awn 3 mm long; palea ovate-elliptic. Caryopsis ovate (Illus. Plate-VI; Fig.-8).

Ecology: Usually found under the shade of trees. Sometimes occurs in agricultural fields. Common.
Fl. & Fr.: July-Oct.
Distribution: Throughout India. Old world tropics.
Specimens examined: Ahri, 1074; Kathodumar, 1049.

2. APLUDA L.

Apluda mutica L., Sp. Pl. 82. 1753; Bor, Grass. Burm. Ceyl. India Pakistan 93. 1960; Cope in Nasir & Ali, Fl. W. Pak. 143: 318. SreeKumar & Nair, Fl. Kerala Grasses 32. 1991; Singh et al., Fl. Bihar 589. 2001; Paria & Chattopadhyay, Fl. Hazaribagh District 2: 1037. 2005; Kabir & Nair, Fl. Tamil Nadu Grasses 359. 2009. *Apluda aristata* Linn.; Amoen. Acad. 4: 303. 1756. *Apluda varia* Hack. in DC., Monogr. Pha. 6: 197. 1889; Hook. f. in Fl. Brit. India 7: 150. 1897; Haines, Bot. Bihar & Orissa pt. 3: 1104. 1961 (Rep. ed.)

Perennials. Culms 0.5-1 m tall, leafy. Leaf blades linear-lanceolate, acuminate at open, glabrous, upper leaves spathiform; sheaths short, tight, glabrous; ligules membranous. Inflorescene of several racemes each enclosed in spathe together forming a leafy panicle. Spikelets in groups of 3, one sessile, bisexual, second imperfect, reduced at the base and a terminal male or neuter, raely bisexual. Sessile spikelets 4-5 mm long. Lower glume 11-13 nerved; upper glume cymbiform, 7-nerved, 2-tipped. Lower lemma 4-5 mm long, folded, hyaline, 3-nerved, awned or awnless. Anthers 3, yellow, ca 1 mm long. Upper lemma 3-nerved, notched, awned (8-10 mm), bent. Second spikelet male. Anthers 3, yellow, ca 2.2 mm long. Terminal spikelet pedicellate, similar to sessile but upper lemma not notched, neuter or male, pedicels ca 3 mm long. Caryopsis oblong, with large embryo (Illus. Plate-VII; Fig.-9).

Ecology: Common under shade, slopes of rocky hills, hedges, bushes and in damp places.

Fl. & Fr.: Sept.-Nov.
Distribution: Throughout India.
Specimens examined: Mainukhar, 7078; Sanjha, 7014.
Uses: As a fodder.

3. ARISTIDA L.

Key to the Species

1a.	Awns of glumes 3-5 mm long	...... 2. *A. setacea*
1b.	Awns of glumes upto 2 mm long	...... 1. *A. adscensionis*

1. Aristida adscensionis L., Sp. Pl. 82. 1753; Hook. f., Fl. Brit. India 7: 224. 1896; Haines, Bot. Bihar & Orissa pt. 5: 977. 1924 (Repr. ed., 3: 1022. 1961); Bor, Common Grasses of the United Provinces 72. 1940; Bor, Grass Burm. Ceyl. Ind. Pak. 407. 1960; Sreekumar & Nair, Fl. Kerala Grasses 326. 1991; Singh et al., Fl. Bihar 590. 2001; Kabir & Nair, Fl. Tamil Nadu Grasses 110. 2009.

Annuals or perennials. Culms 20-60 cm tall, erect or decumbant, tufted. Leaves 10-30 cm long, usually glabrous; ligule a fringe membrane. Inflorescence a contracted panicle; branches filiform. Spikelets 6-9 mm long, straw-coloured or with purple tinge. Lower glume 5-8 mm long, acute, membranous; upper glume 6-9 mm long, 2-dentate at apex. Lemma linear, laterally compressed, ca 8 mm long, scabrid along keel; callus ca 0.5 mm long, pointed, hairy at base; awn 3-fid, not articulated, central one up to 15 mm long, lateral branches slightly shorter. Anthers 3, yellowish purple, ca 1.7 mm long. Caryopsis cylindrical, equalling lemma (Illus. Plate-VIII; Fig.-10).

Ecology: Very Common on sandy gravelly soil, heavly grazed open lands, thin forests, etc.
Fl. & Fr.: Aug.-March.
Distribution: Jharkhand, Bihar, West Bengal, Orissa, etc
Specimens examined: Bukar, 1098; Pathalgarwa, 1134.

2. Aristida setacea Retz., Obs. Bot. 4: 22. 1786; Hook. f., Fl. Brit. India 7: 225. 1896; Prain, Bengal Pl. 2: 1211. 1903; Haines, Bot. Bihar & Orissa pt. 5:

977. 1924; Bor, Grass. Burma Ceylon India & Pakistan 412. 1960; Sreekumar & Nair, Fl. Kerala Grasses 328. 1991; Singh et al. Fl. Bihar 591. 2001; Paria & Chattopadhyay, Fl. Hazaribagh District 2: 1039. 2005; Kabir & Nair, Fl. Tamil Nadu Grasses 114. 2009.

An erect, tufted perennial, 30-80 cm long. Culms slender or stout, erect, glabrous. Leaves 10-20 x 0.2-0.5 cm, filiform or linear, subcoriaceous, smooth; sheath long, smooth; ligule of minute hairs. Panicle 12-25 cm long, contracted; rachis slender, smooth. Spikelets erect; pedicels capillary, scaberulous. Callus densely white bearded. Gl. I 1-1.3 cm long (including the short awn), linear-lanceolate. Gl. II slightly longer than gl. I, awn often 4 mm. Lemma 1.5 cm long, 3-nerved, awn 3-fid, 2.5-3 cm long, scaberulous, Stamens 3 (Plate-XXIX to Plate-XXXII; Fig.-33 to 38 and 40).

Ecology: Subgregarious on sandy and gravelly soils. Common.
Fl. & Fr.: Aug.-Dec.
Distribution: West Bengal, Jharkhand, Bihar, Orissa, U.P., M.P., Western India, Deccan Peninsula. Mayanmar, Sri Lanka to Mascarene Islands.
Specimens examined: Silodhar, 1124; Morainia, 4049.
Uses: Cattles graze young plants. Culms are used for making brooms, brushes, screens, and frames for paper manufacture.

4. **BAMBUSA** Schreb., nom. cons.

Bambusa arundinacea (Retz.) Willd., Sp. Pl. 2: 245. 1799; Gamble in Hook. f., Fl. Brit. India 7: 395. 1896; Haines, Bot. Bihar & Orissa pt. 5: 950. 1924 (Repr. ed., 3: 995. 1961); Singh et al., Fl. Bihar 595. 2001; Paria & Chattopadhyay, Fl. Hazaribagh District 2: 1043. 2005. *Bambos arundinacea* Retz. Observ. Bot 5: 24. 1789.

Local name: Kanta bans, Ketua.

Perennials. Densely tufted bamboo. Culms bright green, shining, branched, 15-35 m tall and 10-15 cm in diam, with to 50 cm long internodes; basal branches horizontal, armed at nodes with 2-3 recurved spines; nodes prominent, lower rooting. Culm-sheaths orange-yellow, with dense golden hairs, usually stripped with green or red; limb triangular, to 10 cm long;

margins decurent on sheaths, wavy, plaited, ciliate, with very narrow acuminate at apex, smooth or scaberulous on margins, glabrous above, hairy beneath, 7-20 cm long and 0.2-1cm broad; sheaths at first hirsute, then glabrescent. Panicle-branches spicate with a few loose clusters of spikelets. Spikelets about 5 in each cluster, lanceolate, sessile, 1.75-2.5 cm long, 4 to 6-flowered. Glumes 2, ovate-lanceolate, mucronate, or absent. Lemmas sometimes shortly fimbriate at top, glabrous. Paleas narrower than lemmas but longer; keels ciliate. Anthers sometimes with an apiculate bristle. Stigmas 2-3, plumose. Caryopsis oblong, 5-8 mm long (Plate-XXXII; Fig.-39 and 40).

Ecology: Common on low hills.
Fl. & Fr.: Once in life.
Distribution: Throughout India except the Himalayas: Mayanmar, Sri Lanka and West Indies.
Specimens examined: Hathia baba, 1044; Garmorwa, 4050.
Uses: Culms are used in construction purposes. These also yield a good quality paper. Young shoots are pickled or made into curries. Leaves and tender part of twigs are used as fodder.

5. BOTHRIOCHLOA O. Kuntze

Key to the Species

1a. Culms more than 100 cm tall; leaves cauline; panicle more than 15 cm long, raceme whorled to semi-whorled ...... 1. *B. bladhii*
1b. Culms less than 100 cm tall; leaves often basal; panicle less than 15 cm long, raceme digitate to subdigitate ...... 2. *B. pertusa*

1. Bothriochloa bladhii (Retz.) S. T. Blake in Proc. Roy. Soc. Queensland 80: 62. 1969; Cope in Nasir & Ali, Fl. W. Pakistan 143: 284. 1982; Sreekumar & Nair, Fl. Kerala Grasses 1: 49. 1991; Singh et al., Fl. Bihar 596. 2001; Paria & Chattopadhyay, Fl. Hazaribagh District 2: 1044. 2005; Kabir & Nair, Fl. Tamil Nadu Grasses 370. 2009. *Andropogon bladhii* Retz., Obs. Bot. 1: 27.1781. *Andropogon intermedius* R. Br. Prodr. 202. 1810; Hook. f., Fl. Brit.

India 7: 175. 1896: Prain, Bengal Pl. 2: 1204. 1903. *Amphilophis glabra* (Roxb.) Stapf in Prain. Fl. Trop. Africa 9: 172. 1917; Haines, Bot. Bihar & Orissa pt. 5: 1028. 1924. *Bothriochloa intermedia* (R. Br.) A. Camus in Ann. Soc. Linn. Lyon 1930, n.s. 76: 164. 1931; Bor, Common Grasses of the United Provinces 81. 1940; Bor, Grass. Burma Ceylon India & Pakistan 108. 1960. *B. glabra* (Roxb.) A. Camus in Ann. Soc. Linn. Lyon 1930, n.s. 76: 164. 1931; Bor, Grass. Burma Ceylon India & Pakistan 107. 1960. *B. caucasica* (Trin.) Hubb. in Kew Bull. 1939. 101. 1939; Bor, Grass. Burma Ceylon India & Pakistan 106. 1960.

Local name: Sandhor

Perennials. Culms upto 110 cm tall, tufted, with short rhizome. Leaves linear, acuminate at apex, hairy at base, scabrid on margins sheaths glabrous; ligule a shallow membrane. Inflorescence a panicle hairy, lower branches shorter. Spikelets in pairs, sessile and pedicellate; pedicel and rachis joints flattened and longitudinally grooved, long ciliate. Sessile spikelets 2-3 mm long; callus short, hairy. Lower glume ca 3.2 mm long, 7-nerved, not pitted on back; upper glume ca 3.5 mm long, 3-nerved. Lower lemma neuter; upper lemma ca 1.2 cm long, hyaline, 1-nerved, with ca 1.2 cm long, scaberulous awn, keeled at centre. Anthers 2, yellow, 1.2-1.4 mm long. Pedicellate spikelets neuter. Caryopsis ca 1.2 mm long.

Ecology: Very common along roadside, forest glades, etc.
Fl. & Fr.: Aug.-Jan.
Distribution: Throughout India.
Specimens examined: Danua, 1028; Chordaha, 1029.
Uses: Young plant is relished well by cattle; culms yield a good paper pulp.

2. Bothriochloa pertusa (L.) A. Camus in Ann. Soc. Linn. Lyon 1930. n.s. 76: 164. 1931; Bor, Common Grasses of the United Provinces 83. 1940; Bor, Grass. Burm. Ceyl. Ind. Pak. 109. 1960; SreeKumar & Nair, Fl. Kerala Grasses 54. 1991; Singh et al., Fl. Bihar 596. 2001; Kabir & Nair, Fl. Tamil Nadu Grasses 373. 2009. *Holcus pertusus* L., Mant. Pl. 301. 1771. *Andropogon pertusus* (L.) Willd., Sp. Pl. 4: 922. 1806; Hook. f., Fl. Brit. India 7: 1173. 1896. *Amphilophis pertusa* (L.) Nash ex Stapf in Agric. News W. Ind. 15: 179. 1916; Haines, Bot Bihar & Orissa pt. 5: 1030. 1924 (Repr. ed., 3: 1077. 1961.)

Perennials. with variable habits. Culms 30-80 cm tall, erect or ascending. Leaves linear, glabrous except at base; sheaths hardly compressed; ligules membranous, truncate. Inflorescence a subdigitate panicle, silky villous. Sessile spikelets 3-mm long; callus bearded. Lower glume ca 3.2 mm long, pitted on back. 5-7-nerved. Lower lemma ca 2 mm long; upper lemma narrowly linear. 10-20 mm long including stipe, awned, kneed and twisted, scaberulous, Anthers 3, yellow, 1.5-2 mm long, sometimes 2 anthers and one staminode (Illus. Plate-IX; Fig.-11 and Plate-XXXIII; Fig.-41 and 42).

Ecology: Along roadside, forest glades, etc.
Fl. & Fr.: Throughout the year if moisture is available.
Distribution: Throughout India.
Specimens examined: Chordaha, 1080; Garmorwa, 1126; Pathalgara, 1138.

6. BRACHIARIA Griseb.

Key to the species

1a.	Spikelets 1.5-2 mm long, secund; pedicels with bulbous-based bristles	...... 2. *B. reptans*
1b.	Spikelets 3-4 mm long, not secund; pedicels without bristles	...... 1. *B. ramosa*

1. Brachiaria ramosa (L.) Stapf in Prain, Fl. Trop. Afr. 9: 542. 1919; Haines, Bot. Bihar & Orissa pt. 5: 1005. 1924 (Repr. ed. 3: 1051. 1961); Bor, Grass. Burm. Ceyl. Ind. Pak. 284. 1960; SreeKumar & Nair, Fl. Kerala Grasses 223. 1991; Singh et al., Fl. Bihar 598. 2001; Paria & Chattopadhyay, Fl. Hazaribagh District 2: 1046. 2005; Kabir & Nair, Fl. Tamil Nadu Grasses 220. 2009. *Panicum ramosum* L., Mant. Pl. 29. 1767; Hook. f., Fl. Brit. India 7: 36. 1896.

Annual. Culms fascicled, suberect or ascending, rooting at base, terete, branched below, to 60 cm long, Leaf-blades linear-lanceolate, rounded at base, flaccid, sharply scabrid on margins, 5-12.5 cm long and 0.6-1.5 cm broad; sheaths thin, loose below, striate, ciliate near mouth; ligule a line of short white hairs. Panicles 5-15 cm long; common axis angular, channeled, scabrid on edges; racemes 5-many, 3-5 mm long, upper gradually shorter;

rachis triquetrous, scabrid. Spikelets ovoid, somewhat turgid, usually paired or fascicled, 2.5-3 mm long. Lower glume broadly ovate, clasping at base. 1.2-1.5 mm long; upper distinctly 7-nerved. Lower floret neuter; lemma similar to upper glume, 5-nerved; palea oblong, ± reduced. Upper floret slightly shorter than lower, ovate; lemma and palea transversely rugose. Caryopsis broadly elliptic, depressed on both sides, to 2 mm long (Plate-XXXIV; Fig.-43).

Ecology: Very common; in cultivated fields, gardens, roadsides, etc.
Fl. & Fr.: June-Dec.
Distribution: Throughout India. Bangladesh, Pakistan, Afganistan, tropical Asia, Senegal to Yemen and southwards to Malawi and South Africa.
Specimens examined: Sanjha, 1008; Khairtanr, 1075.
Use: Plant is grazed by cattle.

2. Brachiaria reptans (L.) Gardner & Hubbard in Hook. Icon. Pl. 34: t. 3363. 1938; Bor, Grasses Burma Ceyl. Ind. Pak. 285. 1960; Cope in Nasir & Ali, Fl. Pakistan 143: 205. 1982; SreeKumar & Nair, Fl. Kerala Grasses 224. 1991; Singh et al., Fl. Bihar 598. 2001; Kabir & Nair, Fl. Tamil Nadu Grasses 222. 2009. *Panicum reptans* L., Syst. Nat. ed. 10. 870. 1759. *Panicum prostratum* Lam. Tab. Encycl, Meth. Bot. 1: 171. 1791; Hook. f., Fl. Brit. India 7: 33. 1896; Prain, Bengal Pl. 2: 1177. 1903 (Rept. ed. 2: 856. 1963). *Urochloa reptans* (L.) Stapf in Fl. Trop. Africa 9: 601. 1920; Haines, Bot. Bihar & Orissa pt. 5: 1003. 1924; Blatter & McCann, Bombay Grasses 144. 1935.

A slender annual. Culms 20-45 cm long, creeping and rooting below. Leaves 4-5.5 x 0.6-1.2 cm ovate-lanceolate, acuminate, glabrous above, base amplexicaul; sheaths striate, with ciliate margins; ligule a tuft of white hairs. Panicle of 5-9 spiciform racemes often 1.2-1.8 cm long; rachis hispidulous, angular. Spikelets 2 mm long, ellipsoid, acute, glabrous; pedicel 0.5-0.75 mm with few long cilia. Gl. I semilunate, 0.5 mm long, hyaline. Gl. II ovate, acute, 7-nerved. Lower lemma empty, elliptic-ovate; palea hyaline, Upper lemma 1.5 mm long; palea subcoriaceous.

Ecology: Common in grazed fields and in deciduous forests etc.
Fl. & Fr.: Aug. - Oct.
Distribution: Throughout India.
Specimens examined: Dhoria, 1013; Muria, 4091.

7. CHLORIS Sw.

Key to the species

1a.	Empty lemma solitary above the fertile floret	...... 2. *C. dolichostachya*
1b.	Empty lemma 2-4 above the fertile floret	...... 1. *C. barbata*

1. Chloris barbata Sw., Fl. Ind. Occ. 1: 200. 1797; Hook. f., Fl. Brit. India 7: 292. 1896; Prain. Bengal Pl. 2: 1228. 1903 (Rept. ed. 2: 926. 1963); Haines, Bot. Bihar & Orissa pt. 5: 969. 1924; Bor, Common Grasses of the United Provinces 89. 1940; Bor, Grasses Burma Ceyl. Ind. Pak. 425. 1960; Cope in Nasir & Ali, Fl. Pakistan 143: 121. 1982; Singh et al., Fl. Bihar 601. 2001; Paria & Chattopadhyay, Fl. Hazaribagh District 2: 1048. 2005. *Andropogon barbatus* L., Mant, 302. 1771 (non. L. 1759). *Chloris inflata* Link, Enum. Pl. Hort. Berol. 1: 105. 1821.

Perennials. Culms 25-75 cm tall, geniculate base. Leaves flat, folded in xeric conditions, glabrous: sheaths compressed, keeled, glabrous; ligule a narrow membrane. Inflorescence of digitate spikes (4-17); rachis scabrous. Spikelets 2.5-3 mm long, 2- seriate, imbricate, subsessile. Lower floret solitary, perfect; upper two floret imperfect; rachilla disarticulates above glumes with a tuft of hairs at base. Lower glume ca 1 mm long, 1-nerved, short-awned; upper glume 2-2.5 mm long. with up to 5 mm long awn. Lower lemma broadly elliptic, 2.5-3 mm long membranous, 3-nerved, often tinged with purple; keel ciliate; awn 2.5-4.5 mm long, bristle like; empty lemmas 2, smaller, truncate, with short awns. Anthers 3, yellow, ca 1.5 mm long. Caryopsis fusiform, ca 1.75 mm long (Illus. Plate-X; Fig.-12 and Plate-XXXIV; Fig.-44).

Ecology: Common along roadsides, on boundary walls, etc.
Fl. & Fr.: Aug.-Jan.
Distribution: Throughout India.
Specimens examined: Kathodumar, 1100; Kabilas, 4048.
Uses: Plant is used as good fodder grass for cattle and horses.

2. Chloris dolichostachya Lagasca. Gen. Sp. Pl. 5. 1816; Bor, Grass Burm. Ceyl. Ind. Pak. 466. 1960; SreeKumar & Nair, Fl. Kerala Grasses 358.

1991; Singh et al., Fl. Bihar 601. 2001; Kabir & Nair, Fl. Tamil Nadu Grasses 188. 2009. *C. incompleta* Roth. Nov. Pl. Sp. 60. 1821; Hook. f., Fl. Brit. India 8: 790. 1896; Haines, Bot. Bihar & Orissa pt. 5: 968. 1924 (Repr. ed., 3: 1013. 1961.); Bor, Common Grasses of the United Provinces 91. 1940.

Perennials. Culms 30-100 cm tall, erect or procumbent; nodes glabrous. Leaves flat, 10-25 cm long, glabrous except near sheath; sheaths compressed, hairy at mouth; ligules hairy. Inflorescence of 4-10, digitate spikes; rachis scabrid. Spikelets 2-seriate, 6.5-7 mm long, subsessile. Florets awned, lower floret perfect and upper imperfect. Lower glume lanceolate, ca 2.5 mm long, membranous; upper glume similar in structure, 5-6 mm long, awned. Lower lemma 5-6 mm long, 3-nerved; awn ca 12 mm long; sterile lemma represented by 8-10 mm long awn. Lodicules 2, small. Anthers 3. Caryopsis brown. oblong-lanceolate, 4-4.25 mm long, subtrigonous.

Ecology: Common in sandy localities, roadsides, etc.
Fl. & Fr.: Sept.-Dec.
Distribution: Throughout India.
Specimens examined: Murtiakalan, 1053; Bukar, 4089.

8. CHRYSOPOGON Trin.

Chrysopogon aciculatus (Retz.) Trin., Fund. Agrost. 188. 1820; Haines, Bot. Bihar & Orissa pt. 5: 1035. 1924; Bor, Common Grasses of the United Provinces 96. 1940; Bor, Grasses Burma Ceyl. Ind. Pak. 115. 1960; Parham in Smith, Fl. Vitiensis Nova 1: 378. 1979; SreeKumar & Nair, Fl. Kerala Grasses 59. 1991; Singh et al., Fl. Bihar 602. 2001; Paria & Chattopadhyay, Fl. Hazaribagh District 2: 1049. 2005; Kabir & Nair, Fl. Tamil Nadu Grasses 380. 2009. *Andropogon aciculatus* Retz. Obs. Bot. 5: 22. 1788; Hook. f., Fl. Birt. India 7: 118. 1896; Prain, Bengal Pl. 2: 1205. 1903 (Rep. ed. 2: 907. 1963).

Local name: Chorant

Perennials. Rhizome creeping, culm base creeping, erect portion of culm upto 48.5 cm high. Leaf - blades 1.8-12.1 x 0.25-0.5 cm, flat. Panicle up to 7.5 cm long, long-peduncled, erect; branches 4-5 nate, fragile, spreading, hairy, scabrid, thickened at the top, articulating obliquely with the bisexual spikelet on only one joint, the same carries the bisexual and sessile and 2 male or neuter pedicelled spikelets, articulation not appendaged. Sessile spikelets linear, 2.4-3.4 x 1.2 - 1.4 mm. Lower glumes 2.9-3.1 x 1.1-1.3 mm, obscurely 2-3 nerved, 2-keeled, linear-subulate involute at base, broadly implicate, towards the top. Upper glume glabrous, keeled, lanceolate, acuminate. Lemma of the lower floret oblong-lanceolate, obtuse or acute retrosely ciliate; of upper floret with up to 6.2 mm long awn. Palea small, glabrous; pedicel slender. Glumes: lower 5-6 x 0.8-1 mm, dorsally convex, Keels ciliate at the tip. Upper glumes 4.8-5.1 x 1.2-1.4 mm, margin retrosely ciliate. Lower floret barran; lemma 4.1-4.3 x 0.6 mm, palea small, cuspidate. Upper floret male; lemma 2.8-3.1 x 0.7-0.9 mm, margin retrosely ciliate; palea up to 1.7 x 0.7 mm (Illus. Plate-XI; Fig.-13 and Plate-XXXIV; Fig.-45).

Ecology: Common along the meadows, village roadsides, banks of the rivers, water courses etc.
Fl. & Fr.: Almost throughout the year but mainly during June to September.
Distribution: Tropical Asia, Australia and Polynesia.
Specimens examined: Danua, 1007; Sanjha, 4047.

9. COIX L.

Coix lachryma-jobi L., Sp. Pl. 972. 1753; Hook. in Hook. f., Fl. Brit. India 7: 100. 1896; Prain, Bengal Pl. 2: 1210. 1903; Haines, For. Fl. Chota Nagpur 564. 1910 & Bot. Bihar & Orissa pt. 6: 1063. 1924; Bor, Grass. Burma Ceylon India & Pakistan 264. 1960; SreeKumar & Nair, Fl Kerala Grasses 207. 1991; Singh et al., Fl. Bihar 604. 2001; Paria & Chattopadhyay, Fl. Hazaribagh District 2: 1052. 2005; Kabir & Nair, Fl. Tamil Nadu Grasses 390. 2009.

Annuals or perennials. Culms 1-2 m tall, robust, branching, spongy, glabrous, polished. Leaves flat, 10-50 cm long, cordate at base; sheaths broad, glabrous; ligule a narrow membrane. Inflorescence of suberect spikes. Male spikelets terminal, 2-3-nate, loosely imbricate on a slender rachis, 8-12 mm long, upto 2 mm broad. Lower glume papery, many-nerved, with winged margins, densely ciliate; upper glume oblong-lanceolate, acuminate, thin, 9-nerved. Lemma oblong-lanceolate, acuminate, membranous, faintly

5-nerved. Palea hyaline, 2-keeled. Female spikelet basal, ovoid, surrounded by the hardened bract forming a head like bony involucres, yellowish white or bluish grey, 6-11 mm long. Lower glume ovate-oblong, acute-papery; upper glume thinner. Lemma equal to spikelet. Anthers 3, orange-yellow, ca 5 mm long. Caryopsis bluish grey, broadly ovoid to globose, 6-10 mm long, smooth, polished (Illus. Plate-XII; Fig.-14).

Ecology: Common in roadsides and along water-lodged areas, rivers, etc.
Fl. & Fr.: Aug.-Jan.
Distribution: Throughout the hotter and damper parts of India.
Specimens examined: Ahri, 7029; Chordaha, 7054.

10. CYMBOPOGON Spreng.

C. martinii (Roxb.) Wats. in Atkins. Gaz. N. W. Prov. Ind. 392. 1882; Bor, Fl. Assam 5: 384 and GBCIP 129; Haines, Bot. Bihar & Oriss pt. 5: 1046. 1924; SreeKumar & Nair, Fl. Kerala Grasses 72. 1991; Singh et al., Fl. Bihar 606. 2001; Paria & Chattopadhyay, Fl. Hazaribagh District 2: 1055. 2005; Kabir & Nair, Fl. Tamil Nadu Grasses 397. 2009. *Andropogon martini* Roxb., Fl. Ind. 1: 280. 1820. *A. schoenanthus* L. var. *martinii* Hook. f. in FBI 7: 204. 1896.

Perennials. Culms 1.5-2 m tall, erect, tufted, glabrous; lower nodes often swollen. Leaves linear-lanceolate, flat, cordate and amplexicaul at base, glabrous; sheaths terete, glabrous; ligules membranous. Inflorescence a compound panicle; spathe 3-6 cm long; spatheole ca 2 cm long, enclosing 3-6 spikelets. Sessile spikelets ovate or ovate-oblong, 3-5 mm long, awned. Lower glume ovate-oblong, obtuse, grooved on the back; upper glume lanceolate, acute; keel winged, serrulate. Lower floret empty. Lemma ca 3.5 mm long, nerveless, ciliate, awned. Palea absent. Upper floret bisexual. Lemma ca 3 mm long, 2-lobed, awned. Pedicellate spikelets male. Lower glume lanceolate-oblong, obtuse; upper glume 3-nerved. Lemma hyaline. Palea absent. Anthers 3, yellow-brown, 2.5-3 mm long. Caryopsis oblong, 2.5-3 mm long.

Ecology: Common in slopes of hills.
Fl. & Fr.: Sept.-Dec.

Distribution: Common in hotter parts of India.
Specimen examined: Silodhar, 7005.
Uses: Source of Palmarosa Oil, used in soaps and cosmetics. It is also used for lumbago and stiff-joints.

11. **CYNODON** Rich. ex Pers., nom. cons.

Cynodon dactylon (L.) Pers., Syn. Pl. 1: 85. 1805; Hook. f., Fl. Brit. India 7: 288. 1896; Prain, Bengal Pl. 2: 1227. 1903 (Rep. ed. 2: 925. 1963); Haines, Bot. Bihar & Orissa pt. 5: 966. 1924; Bor, Common Grasses of the United Provinces 110. 1940; Bor, Grasses Burma Ceyl. Ind. Pak. 469. 1960; SreeKumar & Nair, Fl. Kerala Grasses 360. 1991; Singh et al., Fl. Bihar 607. 2001; Paria & Chattopadhyay, Fl. Hazaribagh District 2: 1059. 2005; Kabir & Nair, Fl. Tamil Nadu Grasses 195. 2009. *Panicum dactylon* L., Sp. Pl. 58. 1753.

Local name: Dub

Perennials. Plant with creeping rhizomes, culms 6-32 cm high. Leaves up to 12 x 0.2 cm, glaucous, usually acuminate, sheaths smooth; ligule as a rim of hairs. Spikes 2-8, digitate, 1.5-5 cm long, green or sometimes purplish-tinged, rachis scaberulous, compressed. Spikelets up to 2.2 mm long. Glumes up to 1.5 mm long, 1-nerved, ovate, acute; keel scaberulous. Lemma about as long as the spikelete, ovate-oblong, margin scabrid. Palea little shorter than the lemma, 2-nerved, linear-oblong, obtuse; keels scabrid. Caryopsis up to 1.3 mm long (Illus. Plate-XIII; Fig.-15 and Plate-XXXV; Fig.-46).

Ecology: Common; forms extensive patches in open grounds.
Fl. & Fr.: All round the year.
Distribution: Pantropic.
Specimens examined: Sikda, 1011; Mainukhar, 1082.
Uses: Leaves are very auspicious and used extensively in all religious festivals. The local people employ the juice of the leaves in healing the cuts.

12. DACTYLOCTENIUM Willd.

Dactyloctenium aegyptium (L.) Beauv., Ess. Agrost. Expl. Pl. 15. 1815; Fischer, Fl. Madras 3: 1840 (1273). 1934; Bor, Common Grasses of the United Provinces 112. 1940; Bor, Grass. burma Ceylon India & Pakistan 489. 1960; SreeKumar & Nair, Fl. Kerala Grasses 367. 1991; Singh et al., Fl. Bihar 608. 2001; Paria & Chattopadhyay, Fl. Hazaribagh District 2: 1060. 2005; Kabir & Nair, Fl. Tamil Nadu Grasses 122-123. 2009. *Cynosurus uegyptius* L., Sp. Pl. 72. 1753. *Eleusine aegyptia* (L.) Desf., Fl. Atlant. 1: 85. 1798; Hook. f., Fl. Brit. India 7: 295. 1896; Prain, Bengal Pl. 2: 1230. 1903; Haines, Bot. Bihar & Orissa pt. 5: 970. 1924.

Annuals. Culms compressed, 10-70 cm long, usually erect or ascending from a creeping base; nodes swollen. Leaves up to 15x0.3 cm, sparsely hairy, margins ciliate; sheaths compressed. Spikes usually 2-5, 1-2 cm long, light olive-grey. Spikelets up to 3.5 mm long. Lower glume up to 2mm long. Upper glume 1.5-2.5 mm long. Lemma 2.8 mm long. Palea with 2 narrow or broadly winged ciliate keels. Anthers 1.5-1.7 mm long. Caryopsis 1.3 x 0.6mm, obovoid-globose, rugose (Plate-XXXV; Fig.-47).

Ecology: Common in sandy grassy places, gardens and by the sides of the roads.
Fl. & Fr.: June & October.
Distribution: Throughout India, Burma, and Sri Lanka as well as warmer regions of the old world.
Specimens examined: Morainia, 1017; Chordaha, 4000; Garmorwa, 4088.
Uses: The seed paste is given to woman after child-birth to clear extra blood and to cure pain.

13. DENDROCALAMUS Nees

Dendrocalamus strictus (Roxb.) Nees in Linnaea 9: 476. 1835; Gamble in Hook. f., Fl. Brit. India 7: 404. 1896; Prain, Bengal Pl. 2: 1234. 1903; Haines, For. Fl. Chota Nagpur 585. 1910 & Bot. Bihar & Orissa pt. 5: 947. 1924; Singh et al., Fl. Bihar 609. 2001; Paria & Chattopadhyay, Fl. Hazaribagh District 2: 1061. 2005. *Bambos stricta* Roxb., Pl. Corom. 1: 58. t. 80. 1798.

Local name: Narbans.

Perennials. A densely tufted bamboo. Culms 6-16 m tall, often 2.5-7.5 cm across, solid or with a central cavity. Lower culm-sheaths 7.5-30 cm long, hairy; ligules narrow oblong, rounded at the base, gradually tapering to subcuspidate tip, pubescent beneath; leaf-sheaths striate, hairy. Panicles large of dense globular head, 0.8-3.3 cm across. Spikelets spinescent, the fertile intermixed with sterile smaller ones. Empty glumes 2 or more, ovate, many-nerved. Lemmas ovate, ending in a sharp spine; palea ovate or obovate, the lower ones 2-keeled. Stamens 6. Lodicules absent.

Ecology: Very common; on hill slopes, hilly forest, etc.
Fl. & Fr.: Oct.-Nov.
Distribution: India: From North India through Punjab to Bihar, Jharkhand, Orissa, West Bengal, Central and South India. Pakistan, Mayanmar. Singapore and Java.
Specimen examined: Garmorwa, 1111.
Uses: Clulms are employed for rafters, battens, baskets, sticks, furniture, fishing rods, etc.; also used for paper-pulp. Leaves are used as fodder.

14. DESMOSTACHYA Stapf

Desmostachya bipinnata (L.) Stapf in Dyer, Fl. Cap. 7: 632. 1900; Bor, Grass. Burm. Ceyl. Ind. Pak. 491. 1960; Singh et al., Fl. Bihar 610. 2001; Kabir & Nair, Fl. Tamil Nadu Grasses 124. 2009. *Briza bipinnata* L., Syst. Nat. ed. 10. 2: 875. 1759. *Eragrostis cynosuroides* (Retz.) P. Beauv., Ess. Agrost. 71, 162. 1812; Hook. f., Fl. Brit. India 7: 324. 1896. *Desmostachya cynosuroides* Stapf ex Haines, Bot. Bihar Orissa pt. 5: 962. 1924.

Local name: Kush.

A stout perennial, branching from the base. Culms 30-100 cm high, erect. Leaves linear to linear-lanceolate, 40-45x0.25-0.5 cm, narrowed into a filiform tip, margins minutely scabrid. Panicles 10-40 cm long, linear or oblong. Spikes crowded, up to 2.5 cm long. Spikelets 2-seriate, secund. Gl. I 0.5-0.7 mm long, mucronate, 1-nerved. Gl. II. 1 mm long. Lemma 1.5 mm long, 3-nerved, keels scabrid. Palea 1.2 mm long. Stamens 3 (Illus. Plate-XIV; Fig.-16).

Ecology: Tolerably common in open grass lands, on dry sandy soils.
Fl. & Fr.: June-Oct.
Distribution: Almost throughout India.
Specimen examined: Ahri, 7010.
Uses: Often used in Hindu ceremonies.

15. DICHANTHIUM Willemet

Key to the Species

1a. Ligules short, fringed. Racemes unequal. Upper glume of sessile spikelets mucronulate ...... 2. *D. ischaemum*

1b. Ligules not fringed. Racemes almost equal. Upper glume of sessile spikelets muticous ...... 1. *D. annulatum*

1. Dichanthium annulatum (Forssk.) Stapf. in Prain, Fl. Trop. Afr. 9: 178. 1917; Hanies, Bot. Bihar & Orissa pt. 5: 1039. 1924 (Repr. ed. 3: 1087. 1961); Bor, Common Grasses of the United Provinces 116. 1940; Bor, Grass. Burm, Ceyl. Ind. Pak. 133. 1960; SreeKumar & Nair, Fl. Kerala Grasses 76. 1991; Singh et al., Fl. Bihar 610. 2001; Paria & Chattopadhyay, Fl. Hazaribagh District 2: 1062. 2005; Kabir & Nair, Fl. Tamil Nadu Grasses 403. 2009. *Andropogon annulatus* Forssk., Fl. Aegypt - Arab. 173. 1775; Hook. f., Fl. Brit. India 7: 196. 1896.

Perennial; culm 0.5-1.4 m high, erect or ascending, simple or less branched, nodes villous. Leaf blades 2.9-24.8 x 0.3-0.5 cm. glabrous or sparsely hairy above; sheaths striate, subcarinate at the top; ligule 2-4 mm long; membranous, oblong, obtuse. Racemes usually 3-12, very rarely solitary or 2, 2.5-7 cm long, pinkish, brownish or greenish; rachis long-hairy, articulations oblique without appendages. Sessile spikelets up to 3-5 mm long, awned. Lower glume 3-5 x 1-1.5 mm, 5-9 nerved; keels spinulosely ciliate. Upper glume 3-5 x 1.1-1.3 mm, lanceolate, muticous. Lower floret; lemma 1.5-2.5 mm long, scabrous, more or less distinctly keeled, with 2 cm long slender awn. Pedicelled spikelets flattened. Lower glume 2.6-2.9 x 1.1-1.5 mm, ciliate. Upper glume 3-nerved. Lower floret: lemma 2.7-3.2 x 0.8-1.2 mm, obtuse, ciliate. Upper floret: lemma muticous, 1.1-2.1 x 0.4-0.6 mm (Illus. Plate-XV; Fig.-17).

Ecology: Generally in thickets, hedges and along roadside.
Fl. & Fr.: Nov.-June.
Distribution: Pantropic.
Specimens examined: Muria, 1150; Silodhar, 4021.

2. Dichanthium ischaemum (L.) Roberty in Boissiera 9: 160. 1960; Deshpande in Jain et al., Fl. India Fasc. 15: 16. 1984; Paria & Chattopadhyay, Fl. Hazaribagh District 2: 1064. 2005. *Andropogon ischaemum* L., Sp. Pl. 1047. 1753; Hook. f., Fl. Brit. India 7: 171. 1897. *Bothriochloa ischaemum* (L.) Keng in Contr. Biol. Lab. Sci. Soc. China, Bot. Ser. 10: 201. 1936; Bor, Grass. Burma Ceylon India & Pakistan 108. 1960.

Culms 40-100 cm long, erect or aciculately ascending; nodes glabrous or shortly bearded. Leaf-sheaths compressed, with long hairs at mouth; blades 6-20 cm long, narrowly linear, globrous or hairy towards base, glabrous, rounded at base; ligule short, fringed. Racemes 3-20, slender, unequal, 6-7 cm long, (sub) digitate; joints and pedicels silky-hairy, ca 2 mm long. Spikelets 3-5 mm long; callus shortly bearded. Sessile spikelets ca 4 mm long, elliptic-lanceolate; lower glume 5-7 nerved, dorsally flat, hairy below middle; upper lanceolate, mucronulate; keels scabrid above; lemma oblong-lanceolate, acute, ciliate; palea similar to lemma, with 1.2-2 cm long awn. Pedicelled spikelets darker than sessile ones, male; lower glume lanceolate, 9-nerved, glabrous; keels ciliate above; upper lemma linear-lanceolate, 5-7 nerved; lemma linear-oblong, obtuse, ciliate, 2.8-3 mm long; palea similar to lemma, but very narrow. Caryopsis 2-2.5 mm long.

Ecology: Common; in waste places, pasture lands, etc.
Fl. & Fr.: Aug.-Dec.
Distribution: India: Jammu and Kashmir, Punjab, Uttar Pradesh, Jharkhand, Rajsthan, Madhya Pradesh, Tamilnadu and Manipur. Pakistan, Afganistan, Africa
Specimens examined: Morainia, 1027; Asnachuan, 4011.

16. DIGITARIA Rich.

Key to the Species

1a. Verrucose hairs present on the spikelets ...... 3. *D. longiflora*
1b. Verrucose hairs absent on the spikelets:
 2a. Spikelets of each pair heteromorphous; lower lemma of sessile spikelet 7-nerved, pectinate; anthers ca 0.5 mm long ...... 1. *D. bicornis*
 2b. Spikelets of each pair not heteromorphous; lower lemma of sessile spikelet 3-nerved, glabrous; anthers ca 1.2 mm long ......2. *D. ciliaris*

1. Digitaria bicornis (Lam.) Roem. & Schult. ex Loud., Syst. Veg. 2: 470. 1817; Bor, Grasses Burma Ceyl. Ind. Pak. 1960; Lazarides, Trop. Grasses S.E. Asia 144. 1980; SreeKumar & Nair, Fl. Kerala Grasses 235. 1991; Singh et al., Fl. Bihar 611. 2001; Kabir & Nair, Fl. Tamil Nadu Grasses 241. 2009. *Paspalum bicorne* Lam, Tab. Encycl. Meth. Bot 1: 176. 1791. *Digitaria barbata* Willd., Enum. Pl. Berol. 91. 1809. *Digitaria rottleri* Roem. et Schult., Syst. Veg. 2: 471. 1817. *Panicum bicorne* (Lam.) Kunth, Enu. Pl. 1: 83. 1833. *Panicum heteranthum* sensu Hook. f., Fl. Brit. India 7: 16. 1896 non Nees & Meyen, 1843.

Annuals. Culms 30-50 cm tall, decumbent. Leaves scabrid, sparsely soft hairy; sheaths glabrous to pilose; ligules membranous, truncate. Inflorescence of 2-5 racemes; rachis winged, serrate. Spikelets 2-nate, lanceolate, 2.75-3.5 mm long, glabrous to slightly pubescent. Lower glume variable, 0.2-0.5 mm long, acute, 2-fid; upper glume 1-2.75 mm long, 3-nerved, pubescent. Sterile lemma equal to spikelets 7-nerved, glabrous or slightly pubescent and pectinate; pubescence usually mixed with bristles. Fertile lemma slightly shorter than the spikelet. Anthers 3, purple, ca 0.5 mm long. Caryopsis yellowish, ca 2.5 mm long (Plate-XXXV; Fig.-48).

Fl. & Fr.: Rainy season
Distribution: Tropical Asia.
Specimens examined: Danua, 1076; Pathalgara, 1078; Garmorwa, 1081.

2. Digitaria ciliaris (Retz.) Koeler, Descr. Gram. Gallia & Germania 27. 1802; Parham in Smith, Fl. Vitiensis Nova 1: 326. 1979; Cope in Nasir &

Ali, Fl. Pakistan 143: 228. 1982; SreeKumar & Nair, Fl. Kerala Grasses 237. 1991; Singh et al., Fl. Bihar 611. 2001; Paria & Chattopadhyay, Fl. Hazaribagh District 2: 1066. 2005; Kabir & Nair, Fl. Tamil Nadu Grasses 242. 2009. *Panicum ciliare* Retz. Obs. Bot. 4: 16. 1786. *Panicum adscendens* Kunth in H.B.K., Nova Gen. Sp. 1: 97. 1816. *Digitaria sanguinalis* (L.) Scop. fa. *ciliaris* (Retz.) Haines, Bot. Bihar & Orissa pt. 5: 1008. 1924. *Digitaria adscendens* (Kunth) Henrard in Blumea 1: 92. 1934; Bor in Webbia 11: 350. 1955 & Grasses Burma Ceyl. Ind. Pak. 298. 1960.

Annual grass, 30-80 cm high. Culms slender, rooting at the lower nodes. Leaves 5-20 x 0.4 - 0.9 cm, linear, hairy from tubercle-bases, acuminate; sheaths sparsely hairy; ligule 1.5-2.5 mm long, membranous. Inflorescence a panicle of 3-10 racemes 4.5-10 cm long, racemes arranged in digitate or subwhorled manner. Spikelets binately arranged, 3-3.5 mm long, elliptic-lanceolate, acute; pedicels unequal, triquetrous. Gl. I a small triangular, nerveless scale, densely villous. GI. II 2- 2.5 mm long, lanceolate, 3-nerved, acute. Lower lemma empty, 5-7-nerved; upper hermaphrodite, coriaceous, oblong-lanceolate, 3-nerved. Stamens 3. Caryopsis 1.6-1.8 mm long (Plate-XXXVI; Fig.-49A and 49B).

Ecology: Common in pastures, hedges and under the shade of the thickets and shrubberies; gregariously grown.
Fl. & Fr.: May to November.
Distribution: Warmer parts of the world.
Specimens examined: Dhoria, 1015; Murtiakalan, 4028.

3. Digitaria longiflora (Retz.) Pers., Syn. Plan 1: 85. 1805; Haines, Bot. Bihar & Orissa pt. 5: 1008. 1924; Henr. Monogr. Gen. *Digitaria* 408. 1950; Bor, Grass. Burma Ceylon India & Pakistan 302. 1960; SreeKumar & Nair, Fl. Kerala Grasses 237. 1991; Singh et al., Fl. Bihar 612. 2001; Paria & Chattopadhyay, Fl. Hazaribagh District 2: 1068. 2005; Kabir & Nair, Fl. Tamil Nadu Grasses 244. 2009. *Paspalum longiflorum* Retz., Obs. Bot. 4: 15. 1786; Hook. f., Fl. Brit. India 7: 302. 1896, p.p. Prain, Bengal Pl. 2: 1181. 1903. *Digitaria preslii* (Kunth) Henr., Monogr. *Digitaria* 589, f. 880. 1950; Bor, Grass. Burma Ceylon India & Pakistan 304. 1960.

Annuals. Culms 10-50 cm tall, with short creeping stolons. Leaves linear, hairy or glabrous; ligules membranous. Inflorescence of 2-3 racemes; rachis flat, winged, glabrous. Spikelets ellipsoid, 1.5-2 mm long. Lower glume absent; upper glume 1.5-1.7 mm long, membranous, covered with verrucose hairs. Lower lemma empty, 5-7 nerved; upper lemma bisexual, elliptic, ca 1.5 mm long. Anthers 3, yellow, ca 0.5 mm long, Caryopsis pale, ellipsoid.

Ecology: Common in open places, roadsides, along border of rice fields, etc.
Fl. & Fr.: July- Dec.
Distribution: Throughout India.
Specimens examined: Sanjha, 1109; Kathodumar, 4035.

17. ECHINOCHLOA P. Beauv.

Echinochloa colona (L.) Link, Hort. Berol. 2: 209. 1833; Haines, Bot. Bihar & Orissa pt. 5: 997. 1924 (Repr. ed., 3: 1043. 1961); Bor, Grass Burm. Ceyl. Ind. Pak. 308. 1960; SreeKumar & Nair, Fl. Kerala Grasses 245. 1991; Singh et al., Fl. Bihar 615. 2001; Paria & Chattopadhyay, Fl. Hazaribagh District 2: 1073. 2005; Kabir & Nair, Fl. Tamil Nadu Grasses 252. 2009. *Panicum colonum* L., Syst. Nat. ed. 10. 2: 870. 1759; Hook. f., Fl. Brit. India 7: 32. 1896.

Tufted, slender annual, 30-60 cm long. Culms glabrous. Leaves 10-18 x 0.6-0.8 cm, linear, acuminate, glabrous; sheaths striate; ligule absent. Panicle 5-11 cm long, erect, narrow; spikes from 6 to many, suberect, 1-2.5 cm long; rachis often hispidulous, angled. Spikelets 2.5-3.25 mm long, ovoid, acute to subcuspidate, usually 4-ranked; pedicel cupular at the tip. Gl. I broadly ovate, shortly cuspicate, 1.75mm long, ciliolate. Gl. II elliptic-ovate, acute, 2.5 mm or more often 5-nerved. Lower lemma male, ovate, acute, 2.5 mm long, 5-7 nerved; palea elliptic-oblong, hyaline. Upper lemma bisexual, 2-2.25 mm, ovate; palea ovate, 2 mm long. Caryopsis 1.6-1.8 cm long (Illus. Plate-XVI; Fig.-18 and Plate-XXXVI to XXXVII; Fig.-50 to 52).

Ecology: Abundant in wet rice fields and near swampy lands.
Fl. & Fr.: July-Oct.
Distribution: Throughout India. Widespread in the Tropics and Subtropics.
Specimens examined: Duragara, 1021; Asnachuan, 4029.
Uses: It is a good fodder grass.

18. ELEUSINE Gaertn.

Key to the Species

1a.	Spikes slender, narrow, 1 to 0.5 cm wide, straight, nearly glabrous at base. Seeds oblong, obtusely trigonous	...... 2. *E. indica*
1b.	Spikes stout, broad, to 1 cm wide, incurved, hairy at base. Seeds globose	...... 1. *E. coracana*

1. Eleusine coracana (L.) Gaertn., Fruct. Sem. Pl. 1: 8, t. 1. f. 11. 1789; Hook. f. Fl. Brit. India 7: 294. 1896; Haines, Bot. Bihar & Orissa pt. 5: 970. 1924(Repr. ed., 3: 1015. 1961); Bor, Grass. Burm. Ceyl. Ind. Pak. 492. 1960; SreeKumar & Nair, Fl. Kerala Grasses 370. 1991; Singh *et al.*, Fl. Bihar 615. 2001; Paria & Chattopadhyay, Fl. Hazaribagh District 2: 1077. 2005; Kabir & Nair, Fl. Tamil Nadu Grasses 127. 2009. *Cynosurus coracanus* L., Syst. Nat. ed. 10, 2: 875. 1759.

Local Name: Marua.

Annual. Culms stout upto 1 m high. Leaves usually far overtopping the stem, 30-60 cm long and 5-6 mm broad; sheaths loose, compressed; ligule of hairs. Spikes 4-7, sometimes a solitary below the whorl, stout, broad, to 1 cm wide, wholly or slightly incurved, hairy at base; rachis of spikes usually pubescent at base, trigonous or flattened on back. Spikelets much congested, awnless, 3-5 flowered. Lemmas broadly ovate, usually with 1 to 2 nerves in margins, to 5 mm long. Caryopsis globose, dark-brown, slightly flattened on one side.

Ecology: Cultivated and also sometimes met with as an escape in paddy fields, waste places, etc.
Fl. & Fr.: Sept.-Dec.
Distribution: Throughout India.
Specimens examined: Morainia, 4034; Silodhar, 4069.
Uses: Its grains are used in cakes, puddlings and in preparation of alcoholic beverage; also useful in biliousness.

2. Eleusine indica (L.) Gaertn., Fruct. 1: 8. 1789; Hook. f., Fl. Brit. India 7: 293. 1896; Prain, Bengal Pl. 2: 1229. 1903; Haines, Bot. Bihar & Orissa pt. 5: 970. 1924; Bor, Common Grasses of the United Provinces 124. 1940; Bor, Grass. Burma Ceylon India & Pakistan 493. 1960; Phillips in Kew Bull. 27: 256. 1972; SreeKumar & Nair, Fl. Kerala Grasses 371. 1991; Singh et al., Fl. Bihar 616. 2001; Paria & Chattopadhyay, Fl. Hazaribagh District 2: 1077. 2005; Kabir & Nair, Fl. Tamil Nadu Grasses 128,129. 2009. *Cynosurus indicus* L., Sp. Pl. 72. 1753.

Local name: Mandla.

Annual, 30-60 cm long, Culms slightly compressed. Leaves linear, acute or subacute, glabrous, margins hispidulous towards the apex; sheaths compressed, margins with few long hairs; ligule membranous. Spikes 3-7, digitate with 1 or 2 situated below the inflorescence. Spikelets often 6-flowered, biseriate, secund. Gl. I ovate-lanceolate, 1.75-2 mm long, 1-nerved. Gl. II elliptic 2.5 mm long, with 5-7 green nerves. Lemma boat-shaped, 3-nerved; palea narrowly lanceolate in outline, 2.75 mm long, 2-nerved. Caryopsis trigonous, reddish-brown (Illus. Plate-XVII; Fig.-19 and Plate-XXXVIII; Fig.-53A and B).

Ecology: Common in wet grounds, lawns and roadsides.
Fl. & Fr.: Aug.-Nov.
Distribution: Throughout India. Tropical and Sub-tropical regions of the world.
Specimens examined: Bukar, 1016; Mainukhar, 1052.
Uses: Culms are used for making hats.

19. ERAGROSTIELLA Bor

Eragrostiella bifaria (Vahl) Bor in Ind. For. 66: 270. 1940 & Grass. Burm. Ceyl. Ind. Pak. 494. 1960; SreeKumar & Nair, Fl. Kerala Grasses 374. 1991; Singh et al., Fl. Bihar 616. 2001; Paria & Chattopadhyay, Fl. Hazaribagh District 2: 1080. 2005; Kabir & Nair, Fl. Tamil Nadu Grasses 130. 2009. *Poa bifaria* Vahl. Symb. Bot. 2: 19. 1791. *Eragrostis bifaria* (Vahl) Wight ex Steud., Syn. Pl. Glumac. 1: 264. 1854; Stapf in Hook. f., Fl. Brit. Inda 7: 325. 1896. *E. coromandeliana* (Koen. ex. Rottl.) Trin. in Mem. Acad. Sci. Petersb. 6, 1:

415. 1830; Stapf in Hook. f., Fl. Brit. India 7: 326. 1896; Haines, Bot. Bihar & Orissa pt. 5: 961. 1924. *Poa coromandeliana* Koen. ex Rottl. in Ges. Natur. Fr. N. Schr. 4: 191. 1803.

Perennials. Culms densely tufted, erect, simple, subcompressed, glabrous, 30-90 cm high. Leaf-blades linear, usually flat, or sometimes complicate, rigid, smooth, 5-7 cm long and to 4 mm borad; sheaths scaberulous, keeled; ligule a ciliate line. Receme 25-30 cm long, erect; rachis smooth, subcompressed; spikelets ovate-oblong, or sometimes linear, spreading, much compressed, olive-grey or green, 0.6-1.5 cm long, 6-40 flowered. Lower glume very acute or acuminate, 3-3.2 mm long, with scaberulous keel; upper 2-2.5 mm long, with stout rounded keel. Lemma broadly ovate, 3-3.5 mm long, palea slightly shorter than glume, with boadly winged scaberulous keel above, faintly nerved. Caryopsis brown, obovoid-ellipsoid, 0.4-0.5 mm long.

Ecology: Common; in dry pasture lands, paddy-fields, rocky places etc.
Fl. & Fr.: Sept.-Dec.
Distribution: From Rajasthan to Bihar, Jharkhand, Orissa, Madhya Pradesh, Deccan Peninsula; Myanmar, Sri Lanka.
Specimens examined: Chordaha, 1097; Pathalgarwa, 4015.
Use: Plant is used as fodder grass for cattle.

20. **ERAGROSTIS** Beauv.

Key to the Species

1a. Spikelets breaking up from above downwards; rachis fragile:
  2a. Panicles effuse. Lemmas not ciliate on margins ...... 5. *E.tenella*
  2b. Panicles spiciform. Lemmas ciliate on margins:
  Spiciform panicles terete. Lemmas mucronate or cuspidately acuminate ...... 2. *E. ciliata*
1b. Spikelets breaking up from below upwards; rachis tough:

3a. Perennials:
  4a. Spikelets 2-2.5mm wide. Lemmas 1.5-1.8 mm long. Paleas deciduous with lemmas ...... 1. *E. atrovirens*
  4b. Spikelets 1-1.2 mm wide. Lemmas 1-1.2 mm long. Paleas persistent ...... 4. *E. nutans*
3b. Annuals:
  5a. Spikelets ovate or ovate-oblong, 2.5-4mm wide, tinged with pale or purple. Rachilla straight, closely nodose ...... 7. *E. unioloides*
  5b. Spikelets linear or linear-oblong, less than 0.2 cm wide, blue-grey or brown. Rachilla zig-zag, with distinct internodes:
    6a. Spikelets more or less fascicled on primary or Secondary branches, all pointing forwards, or shortly pedicelled in narrow racemes ...... 3. *E. gangetica*
    6b. Spikelets not fascicled, long-pedicelled, more or less devaricate when ripe; branches of panicle solitary ...... 6. *E. tremula*

1. Eragrostis atrovirens (Desf.) Trin. ex Steud., Nom. bot. ed. 2, 1: 562. 1840. Bor, Grass. Burma Ceylon India & Pakistan 503. 1960. SreeKumar & Nair, Fl. Kerala Grasses 377. 1991; Singh et al., Fl. Bihar 617. 2001; Paria & Chattopadhyay, Fl. Hazaribagh District 2: 1083. 2005; Kabir & Nair, Fl. Tamil Nadu Grasses 137. 2009. *Poa atrovirens* Desf., Fl. Atlant. 1: 73, t. 14. 1798.

Perennials, with woody base. Culms 50-80 cm tall, erect. Leaves narrow, linear, 10-20 cm long, acute at apex; sheaths striate, smooth; ligules membranous, truncate. Inflorescence a branching panicle, ovate or oblong, 5-20 cm long. Spikelets grey or purplish, 5-10 mm long, 6-12 flowered, glaucous; pedicels 1-3 mm long. Lower glume lanceolate, ca 1.4 mm long, acute, 1-nerved; upper glume cymbiform, lanceolate-acute, ca 1.8 mm long, 1-nerved. Lemma 1.4-2 mm long, acute, 3-nerved, scabrid along mid nerve. Palea narrowly ovate, equaling lemma; keels scabrid. Anthers 3, brownish yellow, sometimes purple, 0.6-1 mm long. Caryopsis reddish brown, cylindrical, ca 0.5 mm long.

Ecology: In watercourses, swamps, banks of streams, etc.
Fl. & Fr.: Aug.-Nov. but in moist places throughout the year.
Distribution: Almost throughout India.
Specimens examined: Morainia, 1054; Muria, 1137.
Use: Plant is eaten by animal.

2. Eragrostis ciliata (Roxb.) Nees, Agrost. Bras. 512. 1829; Stapf in Hook. f., Fl. Brit. India 7: 313. 1896; Haines, Bot. Bihar & Orissa pt. 5: 956. 1924; Bor. Grass. Burm. Ceyl. Ind. Pak. 506. 1960; Singh et al., Fl. Bihar 618. 2001; Paria & Chattopadhyay, Fl. Hazaribagh District 2: 1083. 2005; Kabir & Nair, Fl. Tamil Nadu Grasses 139. 2009. *Poa ciliata* Roxb., Fl. Ind. (Carey & Wallich ed.) 1: 336. 1820.

Perennial. Culms erect or geniculately ascending, terete, 40-250 cm high. Leaf-blades linear-lanceolate, flat or convolute, spreading, acuminate, smooth, glabrous, 5-15 cm long and 2.5-4 mm broad; sheaths bearded below mouth; ligule of a few hairs. Panicle spiciform, cylindric, dense, 5-8 cm long and 4-6 mm in diam.; rachis finely bearded at nodes. Spikelets much compressed, 2.5-4 mm long, 6 to 12-flowered. Glumes ovate, acute, ciliate. Lemma broadly ovate, mucronate or cuspidately acuminate, with minutely scaberulous keel and ciliate margins, 2-3 mm long. Palea shorter than lemma, with strongly reduplicate ciliate keels. Caryopsis ovoid-ellipsoid, terete, 0.5 mm long.

Ecology: On sandy river-beds.
Fl. & Fr.: Oct.-Dec.
Distribution: India: From Maharasthra through southern and central India to Jharkhand and West Bengal. Myanmar and Indo-China.
Specimens examined: Pathalgara, 1039; Duragara, 4075.
Use: Plant is grazed well by cattle.

3. Eragrostis gangetica (Roxb.) Steud. Syn. Pl. Glum. 1: 266. 1854; Haines, Bot. Bihar & Orissa pt. 5: 958. 1924; Bor. Grasses Burm. Ceyl. Ind. Pak. 508. 1960; Lazarides, Trop. Grasses S. E. Asia 174. 1980; SreeKumar & Nair, Fl. Kerala Grasses 380. 1991; Singh et al., Fl. Bihar 618. 2001; Paria & Chattopdhyay, Fl. Hazaribagh District 2: 1085. 2005; Kabir & Nair, Fl. Tamil Nadu Grasses 142. 2009. *Poa gangetica* Roxb. Fl. Indica 1: 340. 1820. *Eragrostis stenophylla* Hochst. ex. Miq. in Verh. Konink. - Nederl. Inst. 3: 4, 39. 1851 pro parte; stapf in Hook. f., Fl., Brit. India 7: 318. 1896. (excl. Syn. *Poa nutans* Retz.); Prain Bengal Pl. 2: 1222. 1903 (Rep. ed. 2: 922. 1963); Haines, Bot. Bihar & Orissa pt. 8: 959. 1924.

Annual. Culms tufted, erect or geniculately ascending, simple or branched, 30-75 cm high; upper internodes long. Leaf-blades narrowly linear, usually convolute, sometimes flat, glabrous, smooth, 7-20 cm long and 1-3 mm broad; sheaths glabrous; ligule a membranous ring. panicle suberect or nodding, 7.5-12 cm long; branches alternate, rather distant, spreading, when ripe, much divided upwards; rachis smooth; pedicels of spikelets filiform. Spikelets linear-oblong, 4-6 mm long and 1-1.2 mm wide, 20-30 flowered, crowded on branches, grey; rachilla flexuous. Lower glume 0.7-0.8 mm long, upper 0.8-1 mm long. Lemmas ovate-oblong, 1-1.2 mm long, with smooth keel. Paleas linear-oblong, with scabrid keels, caducous with glumes. Stamens 2; anthers minute. Caryopsis globose-ellipsoid, reddish-brown, ca 4-5 mm long.

Ecology: Very common; on sandy river beds, along banks of streams, moist pasture lands, etc.
Fl. & Fr.: Sept.-Jan.
Distribution: Throughout India. Nepal. Myanmar and Sri Lanka.
Specimens examined: Dhoria, 1033; Kabilas, 1079.
Use: Plant is eaten by grazing animals.

4. Eragrostis nutans (Retz.) Nees ex Steud., Nomencl. Bot. ed. 2. 1: 563. 1840; Bor. Grass. Burm. Ceyl. Ind. Pak. 511. 1960; SreeKumar & Nair, Fl. Kerala Grasses 385. 1991; Singh et al., Fl. Bihar 620. 2001; Paria & Chattopadhyay, Fl. Hazaribagh District 2: 1087. 2005; Kabir & Nair, Fl. Tamil Nadu Grasses 148. 2009. *Poa nutans* Retz., Observ. Bot. 4: 19 1786. *Eragrostis stenophylla* auct. non. Hochst., 1851: Stapf in Hook. f., Fl. Brit. India 7: 318. 1896, p.p.; Haines, Bot. Bihar & Orissa pt. 5: 959. 1924. *Eragrostis elegantula* Nees ex Steud., Syn. Pl. Glumac. 1: 266. 1854; Stapf in Hook. f., Fl. Brit. India 7: 318. 1896.

Erect tufted perennial, 15-40 cm high. Leaves up to 15.2x0.30 cm, glabrous, convolute, tapering at apex; sheaths glabrous, faintly striate; ligule a narrow membranous ring. Panicle up to 12 cm long; branches rather distant, alternate, much divided upwards, spreading on ripening; rachis slender. Spikelets up to 3.5 mm long on slender panicle, ovate or ovate oblong, sometimes linear-oblong, usually 5-25 flowered, rachilla persistent, flexuous. Glumes ovate-oblong subacute; lower glume 0.6-0.9 x 0.3-0.4 mm; upper 1 x 0.5-0.7 mm. Lemma over-lapping, caducous, ovate-oblong, subacute; keel smooth. Palea slightly shorter than the lemma, keel scabrid. Stamens 3. Caryopsis 0.4-0.6 mm long, brownish red, broadly oblong to subglobose.

Ecology: Less common in moist pastures and along the sides of the rivers.
Fl. & Fr.: Sept. to Jan.
Distribution: Throughout India, Burma, Sri Lanka, Tropical Asia and Africa.
Specimens examined: Silodhar, 1091; Chordaha, 1117; Danua, 4032.

5. Eragrostis tenella (L.) P. Beauv. ex Roem. & Schult., Syst. Veg. 2: 576. 1817; Stapf in Hook. f., Fl. Brit. India 7: 315. 1896; Haines, Bot. Bihar & Orissa pt 5: 956. 1924; Bor, Common Grasses of the United Provinces 128. 1940; Bor, Grass. Burma Ceylon India & Pakistan 513. 1960; SreeKumar & Nair, Fl. Kerala Grasses 390. 1991; Singh et al., Fl. Bihar 620. 2001; Paria & Chattopadhyay, Fl. Hazaribagh District 2: 1088. 2005. *Poa tenella* L., Sp. Pl. 69. 1753. *Poa amabilis* L., Sp. Pl. 68. 1753. *P. plumosa* Retz., Obs. Bot. 4: 20. 1786. *Eragrostis amabilis* (L.) Wight & Arn. ex Hook. & Arn. Bot. Beech. Voy. 251. 1838. *E. tenella* var. *plumose* (Retz.) Stapf in Hook. f., Fl. Brit. India 7: 315. 1896; Prain, Bengal Pl. 2: 1221. 1903; Haines, Bot. Bihar & Orissa pt. 5: 957. 1924.

A tufted annual, 12-50 cm high. Culms slender, glabrous. Leaves 5.5-15 x 0.25-0.5 cm, linear, acuminate, flat, glabrous; sheaths glabrous but hairy near the mouth; ligule a few small hairs. Panicle often 5-10 cm long, oblong and effuse; rachis capillary, bearded at the base, 2-5.5 cm long; rachilla zigzag. Spikelets 3-9 flowered, 1-4.5 mm long, subpyramidal or linear-oblong, compressed, Gl. I 0.5 mm or less, broadly ovate-oblong, obtuse, 1-nerved, ciliolate without. G1. II 0.75 mm, always longer than gl. I. Lemma 1 mm long. ovate-oblong, keels pectinately ciliate. Caryopsis ovoid, polished (Plate-XXXVIII; Fig.-54).

Ecology: Frequent in shady soil, wet places and in open pastures.
Fl. & Fr.: Aug.-Feb.
Distribution: Throughout India.
Specimens examined: Ahri, 1030; Sanjha, 1040; Garmorwa, 1084.
Use: Plant is eaten by Cattle.

6. Eragrostis tremula (Lam.) Hochst. ex Steud., Syn. Pl. Glumac. 1: 269. 1854; Stapf in Hook. f., Fl. Brit. India 7: 320. 1896; Haines, Bot. Bihar & Orissa pt. 5: 960. 1924; Bor, Common Grasses of the United Provinces 134. 1940; Bor, Grass. Burm. Ceyl. Ind. Pak. 514. 1960; SreeKumar & Nair, Fl. Kerala Grasses 392. 1991; Singh et al., Fl. Bihar 621. 2001; Paria & Chattopadhyay, Fl. Hazaribagh District 2: 1090. 2005; Kabir & Nair, Fl. Tamil Nadu Grasses 154. 2009. *Poa tremula* Lam., Tabl. Encycl. 1: 185. 1791. *P. multiflora* Roxb., FI. Ind. (Carey & Wallich ed.) 1: 340. 1820.

Annual. Culms densely tufted, suberect or geniculately ascending, 15-75 cm high. Leaf-blades linear-lanceolate, glabrous or sparsely hairy, rounded at base, eglandular and finely serrate on margin, 2.5-20 cm long; sheaths smooth, glabrous, changing to a mauve colour when dry; ligule a rim of short stiff hairs. Panicle effuse, nodding, lax, to 50 cm long and 30 cm broad; rachis stout, angled, scaberulous; branches solitary, ascending, scaberulous; branchlets capillary, hairy in axils; pedicels shorter or longer than spikelets. Spikelets linear, compressed, glabrous, green or tinged with violet, to 2.5 cm long, to 60-flowered. Glumes subequal, 1.2- 1.5 mm long, ovate, 1-nerved, with scaberulous keels. Lemmas broadly ovate, 1.5-2 mm long; paleas obovate-oblong, persistent, with scabrid keels. Stamens 3. Lodicules 2. Caryopsis globose, 0.3-0.4 mm across.

Ecology: Common; in dry sandy places; river-beds, etc.
Fl. & Fr.: Sept.-Dec.
Distribution: Almost throughout India. Myanmar, tropical Africa.
Specimens examined: Kathodumar, 1147; Bukar, 4065.
Uses: Plant is eaten as fodder. It forms a good sand binder. Grains are eaten in times of scarcity.

7. Eragrostis unioloides (Retz.) Nees ex Steud., Syn. Pl. Glum. 1: 264. 1854; Bor, Common Grasses of the United Provinces 129. 1940; Bor, Grass. Burm. Ceyl. Ind. Pak. 515. 1960; SreeKumar & Nair, Fl. Kerala Grasses 392. 1991; Singh et al., Fl. Bihar 621. 2001; Paria & Chattopadhyay, Fl. Hazaribagh District 2: 1091. 2005; Kabir & Nair, Fl. Tamil Nadu Grasses 155. 2009. *Poa unioloides* Retz., Observ. Bot. 5. 19. 1789. *Eragrostis amabilis* Stapf in Hook. f., Fl. Brit. India 7: 317. 1896, non Wight & Arnott ex Nees, 1838; Haines, Bot. Bihar & Orissa pt. 5: 958. 1924 (Repr. ed., 3: 1003. 1961).

Tufted annual, up to 40 cm high. Culms slender, erect or geniculately ascending, glabrous. Leaves 7.5-15 x 0.25-0.6 cm, linear, acuminate, glabrous, base subcordate: sheath clasping, with a few cilia near the mouth; ligule a minute, narrow rim. Panicle 6-12 cm long, ovoid or ovoid-oblong, effuse. Spikelets 0.4-1 x 0.25-0.35 mm ovate-oblong, deep pink, compressed; pedicel short, glabrous; rachilla zigzag, glabrous. Glumes fall off with the lemmas, keeled and spreading, 1-nerved. Gl. I 1.25-2 mm long, ovate, acute. Gl. II 1.8-2.2 mm long, ovate. Lemma 1.5-1.75 mm long, elliptic-ovate, acute; palea 1.75 mm long, obtuse, scabrid, Stamens 3. Caryopsis 0.5-0.75 mm long, ellipsoid, brown (Illus. Plate-XVIII; Fig.-20).

Ecology: Commonly grows in wet ground or near marshy places.
Fl. & Fr.: Oct.-Jan.
Distribution: Throughout India. South-East Asia.
Specimens examined: Pathalgarwa, 1024; Khairtanr, 4025.
Uses: Plant is grazed well by cattle and horses; also used as green manure.

21. ERIOCHLOA H. B. K.

Eriochloa procera (Retz.) C.E. Hubb. in Bull. Misc. Inform. Kew 1930: 256. 1930; Bor, Grass. Burm. Ceyl. Ind. Pak. 312. 1960; SreeKumar & Nair, Fl. Kerala Grasses 250. 1991; Singh et al., Fl. Bihar 622. 2001; Paria & Chattopadhyay, Fl. Hazaribagh District 2: 1092. 2005; Kabir & Nair, Fl. Tamil Nadu Grasses 256. 2009. *Agrostis procera* Retz. Observ. Bot. 4: 19. 1786. *Eriochloa polystachya* Hook. f., Fl. Brit. India 7: 20. 1896, non Humb., Bonpl. & Kunth, 1816. *E. ramosa* Kuntze, Revis. Gen. Pl. 2: 775. 1891; Haines, Bot. Bihar Orissa pt. 5: 1006. 1924 (Repr. ed., 3: 1052. 1960).

Perennial. Culms densly tufted, erect or decumbent at base, usually geniculate, simple or branched, striate, swollen and pubescent at nodes, 0.3-1.5 cm tall. Leaf-blades linear, flat, acuminate, 9.5-25 cm long; sheaths open, compressed, subcarinate, striate; ligule short, hairy. Racemes 5-13 cm long, loosely arranged; rachis angular, smooth; pedicels triquetrous, swollen at apex, silky-hairy. Spikelets laxly imbricate, ovate-lanceolate, velvety, 2.5-3.5 mm long. Lower glume absent; upper ovate-lanceolate, convex, with inrolled margins, silky, 2.5-3.5 mm long. Lower floret sterile; lemma 2.3 -3.2 mm long, epaleate. Upper floret: lemma subcoriaceous, oblong, pale, compressed dorsally with narrowly inflexed margins, hairy at top; palea oblong. Caryopsis lenticular, smooth, ca 2 mm.

Ecology: Common; in damp places, ditches, edges of paddy fields, etc.
Fl. & Fr.: Aug.-Dec.
Distribution: India: From Punjab through central part to Jharkhand, Orissa and southwards. Tropical Africa, throughout S.E. Asia; introduced in tropical America.
Specimens examined: Asnachuan, 1026; Sikda, 1065.
Use: Plant yields a good fodder, relished by the cattle.

22. HACKELOCHLOA O. Kuntze, nom. cons. prop.

Hackelochloa granularis (L.) O. Kuntze, Rev. Gen. Pl. 776. 1891; Raizada et al. in Ind. For. Rec. (N.S.) Bot. 4: 205, 1957; Bor, Grass. Burma Ceylon India & Pakistan 159. 1960; SreeKumar & Nair, Fl. Kerala Grasses 111. 1991; Paria & Chattopdhyay, Fl. Hazaribagh District 2: 1093. 2005; Kabir & Nair, Fl. Tamil Nadu Grasses 425. 2009. *Cenchrus granularis* L., Mant. 2, App. 575. 1771. *Manisuris granularis* (L.) L. f., Nov. Gram. Gen. 40. 1779; Hook. f., Fl. Brit. India 7: 159. 1896; Prain, Bengal Pl. 2: 1191. 1903; Haines, Bot. Bihar & Orissa pt. 5: 1105. 1924

Local name: Trinpali, Kangni.

Annual, up to 75 cm high. Culms slender, compressed, hairy. Leaves 4-15 x 0.4-1.2 cm, linear-lanceolate, acute to subacuminate, hairy, base cordate; sheaths often 2-4 cm, hirsute; ligule membranous, ciliate, Racemes 0.8-2 cm long; linear, solitary or few in the axils. Sessile spikelets 1.5 mm long, globose; callus small. Gl. I subglobose, pitted and tubercled. Gl. II elliptic-oblong, 3-nerved, obtuse, glabrous. Lower lemma 0.5-0.75 mm long, hyaline. Upper lemma 0.5-1 mm, bisexual. Palea 0.5 mm or less, ovate. Pedicelled spikelets 1.5-1.75 mm long. Gl. I ovate, subacute, 5-nerved, Gl. II 1.5 mm long, cymbiform, 5-7 nerved, winged, scabrid. Lower lemma ciliolate, barren. Upper lemma laceolate, hyaline, acute; palea 1-1.25 mm, lanceolate.

Ecology: Grows commonly in open pasture of the forests. It prefers moist and open place.
Fl. & Fr.: Aug.-Nov.
Distribution: Throughout the hotter parts of India. Widespread in the tropics.
Specimens examined: Kabilas, 1119; Morainia, 4001.

23. HETEROPOGON Pers.

Heteropogon contortus (L.) P. Beauv. ex Roem & Schult., Syst. Veg. 2: 836. 1817; Haines, For. Fl. Chotanagpur 579. 1910 & Bot. Bihar & Orissa pt 5: 1040. 1924; Bor, Common Grasses of the United Provinces 142. 1940; Bor, Grass. Burma Ceyon India & Pakistan 159. 1960; SreeKumar & Nair, Fl. Kerala Grasses 112. 1991; Singh et al., Fl. Bihar 624. 2001; Paria & Chattopadhyay, Fl. Hazaribagh District 2: 1094. 2005; Kabir & Nair, Fl. Tamil Nadu Grasses 427. 2009. *Andropogon contortus* L., Sp. Pl. 1045. 1753; Roxb., Fl. Ind. 1: 253. 1820; Hook. f., Fl. Brit. India 7: 199. 1896; Prain, Bengal Pl. 2: 1205. 1903.

Local name: Kher, Kumeria.

A densely tufted perennial, 30-90 cm high. Culms slender, erect or decumbent below. Leaves 10-30 x 0.3-0.55 cm, linear, shortly acuminate, often sparsely ciliate towads the base; sheath keeled, ciliate at the top; ligule short, ciliolate, Racemes 3-7 cm long with every short internodes; spikelets subsecund, lower pairs of 4-8 male or neuter, awnless; upper spikelets female. Sessile (female) spikelets 6 mm long. Callus bearded. Gl. I 4-6 mm long, linear-oblong. Gl. II 4.5-5 mm long, linear, obtuse. Lower lemma oblong, truncate. Upper lemma represented by 7.5 cm long subulate awn. Pedicelled spikelets 0.8-1.2 cm long. Gl. I lanceolate, subacute, densely hispid outside, margins unequally winged. Gl. II 0.9-1.2 cm long, 5-nerved. Lower lemma 6-7 mm long, oblong, truncate. Upper lemma 6-7 mm long, hyaline, Ciliate (Illus. Plate-XIX; Fig.-21 and Plate-XXIX; Fig.-55).

Ecology: Chiefly grows in dry areas of the hills and valleys. Also found in open grounds of the forests.
Fl. & Fr.: Sept.-Jan.
Distribution: Throughout India. Wide spread in other tropical countries.

Specimens examined: Murtiakalan, 1102; Sikda, 4033.
Uses: Plant is used for silage and hay, also used in manufacture of paper. Roots are used as stimulant and diuretic.

24. HORDEUM L.

Hordeum vulgare L., Sp. Pl. 84. 1753; Hook. f., Fl. Brit. India 7: 371. 1896; Haines, Bot. Bihar & Orissa pt. 5: 964. 1924 (Repr. ed., 1009. 1961); Bor, Grass. Burm. Ceyl. Ind. Pak. 677. 1960; Singh et al., Fl. Bihar 625. 2001; Kabir & Nair, Fl. Tamil Nadu Grasses 94. 2009.

Local name: Jau.

Annuals. Culms 30-75 cm tall, erect, tufted. Leaves 15-30 cm long, soft, with well developed auricles; sheaths glabrous, loose; ligules 1.5-2 mm long, membranous. Inflorescence a terminal spike. Spikelets 1-flowered, 3 at each node, falling together at maturity, all spikelets fertile, the triads arranged in 6 longitudinal rows of fertile spikelets in each spike. Glumes linear-lanceolate, 5-6 mm long, hairy, awned. Lemma 7-10 mm long, 5-nerved, hairy on keel; awn 5-7 mm long, scabrid. Anthers 3, pale yellow, linear, 3-3.5 mm long. Caryopsis cream-coloured, 8-9 mm long.

Ecology: Cultivated throughout the state.
Fl. & Fr.: Jan.-March.
Distribution: Throughout India.
Specimens examined: Ahri, 4085; Garmorwa, 4014.
Uses: Grains are used in the form of aata & sattu. It is also used as a coolent.

25. IMPERATA Cyrill.

Imperata cylindrica (L.) P. Beauv. var. **major** (Nees) Hubb. ex Hubb. & Vaughan, Grass. Maur. 96. 1940; Bor, Common Grasses of the United Provinces 144. 1940; Bor, Grass. Burm. Ceyl. Ind. Pak. 170. 1960; SreeKumar & Nair, Fl. Kerala Grasses 114. 1991; Singh et al., Fl. Bihar 626. 2001; Paria & Chattopadhyay, Fl. Hazaribagh District 2: 1096. 2005; Kabir & Nair, Fl. Tamil Nadu Grasses 429. 2009. *I. koenigii* (Retz.) P. Beauv. var. *major* Nees.

Fl. Afr. III. Austral. 90. 1841. *I. arundinacea* Cyrill., Pl. Rar. Neapol. 2: 27. 1792; Hook. f., Fl. Brit. India 7: 106. 1896; p.p.; Prain, Bengal Pl. 2: 1188. 1903, p.p.; Haines, For. Fl. Chotanagpur 569. 1910 & Bihar & Orissa pt. 5: 1015. 1924, p.p.

An erect perennial, 30-75 cm high. Culms solid, nodes glabrous or bearded. Leaves 15-45 x 0.3-0.7 cm, linear, acuminate, flat, glabrous with scabrid margins; sheaths loose, glabrous; ligule short, hairy. Panicle often 6-15 cm long, spiciform, subcylindric, soft, silvery-white. Spikelets 3 mm long, linear-lanceolate; pedicel 1-1.25 mm long, glabrous; callus hairs many, twice as long as the glumes. Gl. I ovate-lanceolate, 3mm long, 5-nerved. Gl. II oblong-lanceolate, 2.75 mm long, obtuse. Lower lemma oblong-lanceolate, 1.75 mm long, hyaline, subacute, apex ciliolate, nerveless. Upper lemma bisexual, lanceolate. Stigmas 2, purple.

Ecology: Frequent in open grounds, sandy river-beds, forest-edges, roadsides, etc.
Fl. & Fr.: Mar. - June and Oct.-Dec.
Distribution: Hotter parts of India, Pakistan, throughout Old World Tropics, extending to Mediterranean and Middle East; also in Chili.
Specimens examined: Pathalgarwa, 1125; Duragara, 1113.
Uses: Plant is used in manufacture of paper. Culms are used in making ropes, brushes, mats, etc. It is also employed as packing materials.

26. ISCHAEMUM L.

Ischaemum indicum (Houtt.) Merr. in J. Arnold Arb. 19: 320. 1938; Bor, Grass. Burma Ceylon India & Pakistan 180. 1960; SreeKumar & Nair, Fl. Kerala Grasses 9: 136. 1991; Singh et al., Fl. Bihar 628. 2001; Paria & Chattopadhyay, Fl. Hazaribagh District 2: 1100. 2005; Kabir & Nair, Fl. Tamil Nadu Grasses 436. 2009. *Phleum indicum* Houtt., Nat. Hist. II, 13: 198, t. 90. f. 2. 1782. *Ischaemum ciliare* Retz., Obs. Bot. 6: 36. 1791; Hook. f., Fl. Brit. India 7: 133. 1896; Prain, Bengal Pl. 2: 1196. 1903; Haines, For. Fl. Chotanagpur 567. 1910 & Bot. Bihar & Orissa pt. 5: 1022. 1924. *Ischaemum aristatum* auct. non L.; Hook. f., Fl. Brit. India 7: 126. 1896, p.p., excl. *I imberbe* Retz.

Local name: Kander.

Perennials. Culms 30-60 cm tall, erect or decumbent. Leaves 10-20 cm long, pubescent; sheaths glabrous; ligules membranous, truncate. Inflorescence of 2-3 spikes. Sessile spikelets oblong-lanceolate, 5-6 mm long. Lower glume ca 5.2 mm long, glabrous; upper glume ca 5 mm long; keels scabrid; awn ca 2 mm long. Lower lemma hyaline, ca 4.2 mm long, glabrous. Palea linear-lanceolate, ca 2 mm long; keels scabrid. Anthers 3, yellow, ca 2 mm long; keels scabrid. Anthers 3, yellow, ca 2 mm long. Upper lemma deeply notched, 3-nerved; awn ca 1.8 cm long. Caryopsis ca 1.5 mm long with persistent style.

Ecology: Common in rice-fields.
Fl. & Fr.: Aug.-Dec.
Distribution: Throughout India.
Specimens examined: Kathodumar, 1135; Silodhar, 4002.
Uses: Plant is eaten by cattle.

27. ISEILEMA Anders.

Key to the Species

1a.	Tubercles present on spathe and/or margins of the lower glume of involucral spikelets	...... 2. *I. prostratum*
1b.	Tubercles absent on spathe and/or margins of the lower glume of involucral spikelets	...... 1. *I. anthephoroides*

1. Iseilema anthephoroides Hack. in A. & C. DC., Monogr. Phan. 6: 683. 1889; Hook. f., Fl. Brit. India 7: 219. 1896; Haines, Bot. Bihar & Orissa pt. 5: 1054. 1924 (Repr. ed., 3: 1101. 1961); Bor, Grass. Burm. Ceyl. Ind. Pak. 187. 1960; Singh et al., Fl. Bihar 629. 2001; Kabir & Nair, Fl. Tamil Nadu Grasses 445-446. 2009.

Annuals. Culms 20-100 cm tall, tufted. Leaves flat, glabrous or sparsely hairy; sheaths compressed, glabrous; ligules lacerate. Inflorescence a narrow panicle; spathes with tubercle-based hairs. Involucral spikelets lanceolate, 6.5-7 mm long, 3-5-nerved. Lemma absent. Pedicellate spikelets reduced to a long pedicel. Bisexual spikelets lanceolate, 6-6.5 mm long, awned. Lower glume 6-6.5 mm long, 2-fid; upper glume similar. Lower lemma empty; upper lemma reduced to 12-14 mm long awn. Anthers 3, yellow, ca 2.5 mm long, Caryopsis not seen.

Ecology: Common on damp places and fields.
Fl. & Fr.: July-Dec.
Distribution: Throughout India.
Specimens examined: Muria, 1148; Dhoria, 4023.

2. Iseilema prostratum (L.) Anderss. in Nov. Act. Soc. Sci. Upsal. ser. 3, 2: 251. 1856; Bor, Grass. Burm. Ceyl. Ind. Pak. 188. 1960; Singh et al., Fl. Bihar 630. 2001; Paria & Chattopadhyay, Fl. Hazaribagh District 2: 1102. 2005; Kabir & Nair, Fl. Tamil Nadu Grasses 449. 2009. *Andropogon prostratus* L., Mant. Pl. 304. 1771. *Iseilema wightii* Anderss. in Nov. Act. Soc. Sci. Upsal. ser. 3, 2: 251. 1856; Hook. f., Fl. Brit. India 7: 218. 1896; Haines, Bot. Bihar & Orissa pt. 5: 1055. 1924 (Repr. ed., 3: 1103. 1961).

Perennials. Culms usually prostrate and rooting at base, finally ascending, branched below, 20-80 cm long. Leaf-blades linear, rounded at base, 6-12 cm long and 2-3 mm broad, scabrid above; sheaths loose, bearded at nodes, compressed, keeled, shorter than internodes; ligule short, ciliate. Panicle oblong, dense, spreading, decompound; rachis with bearded nodes; proper sheaths 0.8-1.3 cm long, linear-lanceolate. Involucral spikelets 3-3.5 mm long, compressed, lanceolate; pedicel short, bearded at base; lower glume 5-nerved, usually verrucose; upper glumes glabrous, 1-nerved, with inflexed ciliate margins. Lower floret absent. Upper floret male; lemma 2-3-toothed, nerveless. Pedicellate spikelets male or neuter. Sessile spikelets hermaphrodite, lanceolate, 1.5-3 mm long; lower glume with inflexed margins and scaberulous keels; upper glume oblong-lanceolate, obliquely carinate, with scabrid keels; lower floret neuter, lemma linear; upper floret hermaphrodite; lemma epaleate.

Ecology: Common; on damp places, fields and field-borders.
Fl. & Fr.: Aug.-Nov.
Distribution: Throughout India. Myanmar.
Specimens examined: Asnachuan, 1115; Mainukhar, 5000.
Use: Plant is used as fodder, preferably as hay.

28. LEERSIA Sw.

Leersia hexandra Sw., Prodr. Veg. Ind. Occ. 21. 1788; Hook. f., Fl. Brit. India 7: 94. 1896; Prain, Bengal Pl. 2: 1185. 1903 (Rep. ed., 2: 892. 1963); Haines, Bot. Bihar & Orissa pt. 5: 981. 1924; Bor, Grasses Burma, Ceyl. Ind. Pak. 599. 1960; Lazarides, Trop. Grasses S.E. Asia 183. 1980; SreeKumar & Nair, Fl. Kerala Grasses 430. 1991; Singh et al., Fl. Bihar 630. 2001; Paria & Chattopadhyay, Fl. Hazaribagh District 2: 1103. 2005; Kabir & Nair, Fl. Tamil Nadu Grasses 42. 2009. *Homalacenchrus hexandrus* (Sw.) O. Kuntze, Rev. Gen. Pl. 777. 1891.

Local name: Jangali dhan.

A weak perennial grass, Culm 60-90 cm long, rooting in the mud, smooth, striate; nodes hairy. Leaves 10-20 x 0.3-0.6 cm, linear, acuminate, narrowed at the base, glabrous, scaberulous on the margins; sheath glabrous; ligule truncate, short. Panicle 5-10 cm long, pedunculate with slender, distant and flexuous branches 1.3-7 cm long. Spikelets 3-3.25 mm long, closely imbricate. Glumes reduced to an obscure 2- lobed rim. Lemma obliquely oblong, strongly keeled, bristly; palea 3-3.25 mm long, linear-lanceolate, keel bristly. Stamens 6.

Ecology: Occurs at the edges of tanks, lakes and in marshy places. Occasional.
Fl. & Fr.: Sept.-Dec.
Distribution: Throughout India. Tropics of the Old and New world.
Specimens examined: Kenduadih, 1141; Sanjha, 5035.
Uses: It affords good forage when cut early. Plant is also used as green feed and hey.

29. LEPTOCHLOA Beauv.

Leptochloa panicea (Retz.) Ohwi in Bot. Mag. Tokyo 55: 311. 1941; Bor, Grasses Burma Ceyl. Ind. Pak. 517. 1960; Cope in Nasir & Ali, Fl. Pakistan 143: 74. 1982; Singh et al., Fl. Bihar 630. 2001; Paria & Chattopadhyay, Fl. Hazaribagh District 2: 1104. 2005; Kabir & Nair, Fl. Tamil Nadu Grasses 160. 2009. *Poa panicea* Retz., Obs. Bot. 3: 11. 1783. *Aira filiformis* Koen. ex. Roxb., Fl. Indica 1: 328. 1820. *Leptochloa filiformis* auct. non Roem. & Schult.;

Hook. f., Fl. Brit. India 7: 1896, p.p.; Prain, Bengal Pl. 2: 1225. 1903; Haines, Bot. Bihar & Orissa pt. 5: 972. 1924.

Annuals Culms 25-60 cm tall, geniculate. Leaves linear-lanceolate with tubercle-based hairs; sheaths glabrous or hairy along margins; ligules lacerate. Inflorescence a panicle of several slender, ascending branches, often purplish. Spikelets usually 2-3 flowered, 1.8-2.5 mm long, overlapping. Lower glume lanceolate, 0.7-1.5mm long, acute; upper glume narrowly oblong, 0.9-1.6 mm long, mucronate. Lemma elliptic or oblong, 0.8-1.2 mm long, 3-nerved. Anthers very small, pinkish. Caryopsis broadly ellipsoid, trigonous, ca 0.5 mm long.

Ecology: Common in moist grassy places.
Fl. & Fr.: June to Oct.
Distribution: Throughout India, Burma, Sri Lanka and all countries of Tropical Asia and Africa.
Specimens examined: Chordaha, 1014; Pathalgara, 5099.

30. OPLISMENUS Beauv.

Oplismenus burmannii (Retz.) P. Beauv. Ess. Agrost. 54, 168, 169. 1812; FBI 7: 68; Bor in Fl. Assam 5: 263 & GBCIP 317; BP 2: 883; Haines, Bot. Bihar & Orissa pt. 5: 999. 1924; SreeKumar & Nair, Fl. Kerala Grasses 256. 1991; Singh et al., Fl. Bihar 634. 2001; Paria & Chattopadhyay, Fl. Hazaribagh District 2: 1105. 2005; Kabir & Nair, Fl. Tamil Nadu Grasses 263. 2009. *Panicum burmannii* Retz. Obs. Bot. 3: 10. 1783.

Annuals. Culms 15-38 cm long, ascending from the creeping base, branched upwards, basal nodes rooting. Leaves 1.5-4.2 x 0.4-0.9 cm, sparsely ciliate, lanceolate, acuminate; sheaths somewhat loose, striate, margin ciliate, ligule small. Inflorescence 3-9 cm long; rachis usually terminated by a spikelet, dorsally flat, fringed with pubescence; pedicels 2-nate or solitary. Spikelets 2.3-2.8 x 0.7-0.9 mm, contiguous, oblong-lanceolate. Glume awned, ciliolate, or pubescent; lower 1.7 x 0.5-0.8 mm, 3-nerved; upper slightly larger than the lower, elliptic-lanceolate, 5-7 nerved, margin softly hairy. Lower floret: lemma 2.1-2.3 x 1.2-1.4 mm, oblong-lanceolate, 7-nerved. Palea O. Upper floret: 2.4-2.6 mm long, brownish, polished. Lemma 1.5-1.8 x 0.8-1.1 mm, papery, apex pointed. Stigmas red. Caryopsis 2.1-2.4 mm long, lanceolate-oblong, convex on the back (Plate-XXXIX; Fig.-56).

Ecology: Common in damp shady places; occasionally grows in association.
Fl. & Fr.: Sept. to Dec.
Distribution: Plains of India, Sri Lanka. Malaya, Islands, China, Japan and Tropical Africa, Tropical America, Asia
Specimens examined: Sikda, 1089; Asnachuan, 5011.

31. ORYZA L.

Key to the Species

1a.	Spikelets persistent, awned or not	...... 2. *O. sativa*
1b.	Spikelets deciduous with age always long-awned	...... 1. *O. rufipogon*

1. Oryza rufipogon Griff., Notul. 3: 5. pl. 144. f. 2. 1851; Bor, Grasses Burma Ceyl. Ind. Pak. 605. 1960; Lazarides, Trop. Grasses S.E. Asia 184. 1980; SreeKumar & Nair, Fl. Kerala Grasses 432. 1991; Singh et al., Fl. Bihar 635. 2001; Paria & Chattopadhyay, Fl. Hazaribagh District 2: 635. 2005; Kabir & Nair, Fl. Tamil Nadu Grasses 44. 2009. *Oryza fatua* Koen. ex Trin., Mem. acad. Sc. Petersb. ser. 6, 3: 177. 1839 nom. nud. *Oryza sativa* L. var. *fatua* Prain, Bengal Pl. 2: 1184. 1903 (Rep. ed. 2: 891. 1963); Chatterjee in Nature 160: 234. 1947 & in Indian J. Ag. Sci 18: 190. 1948. *Oryza nivara* Sharma & Shastry in Indian J. Gen. Pl. Breed. 25: 161. 1965.

Local name: Deodhan.

Annuals. Culms 1-1.5 m tall, lower parts spongy. Leaves linear, acuminate at apex, scabrid on margins and veins; sheaths terete, loose, glabrous, auricled; ligules up to 1.7 cm long, splitting at tip. Inflorescence an exserted, compound panicle; main branches angular and scabrid. Spikelets 8-9 mm long, falling with age. Lemma I 1-nerved, ca 2.5 mm long; lemma II ca 2.4 mm long; lemm III ca 0.7 mm long, strongly folded above the mid nerve; awn ca 7 cm long, scabrid. Palea similar; awn ca 6 mm long. Anthers 6, yellow, ca 5 mm long. Caryopsis 4-5 mm long, similar to *Oryza sativa*.

Ecology: Common in ditches, swamps, paddy fields, etc.
Fl. & Fr.: Sept.-Nov.
Distribution: India; Eastern and Central India and Tamil Nadu. Pantropic.
Specimens examined: Danua, 1101; Chordaha, 1105; Bukar, 1106; Pathalgarwa, 1136.
Uses: Grain is edible and is eaten in the time of scarcity. Young plant is eaten by buffaloes.

2. Oryza sativa L., Sp. Pl. 333. 1753; Hook. f., Fl. Brit. India 7: 92. 1896, p.p; Prain, Bengal Pl. 2: 1184. 1903; Haines, Bot. Bihar & Orissa pt. 5: 980. 1924, p.p.; Bor, Grass. Burma Ceylon India & Pakistan 605. 1960; SreeKumar & Nair, Fl. Kerala Grasses 433. 1991; Singh et al., Fl. Bihar 636. 2001; Paria & Chattopadhyay, Fl. Hazaribagh District 2: 1108. 2005; Kabir & Nair, Fl. Tamil Nadu Grasses 46. 2009.

Local name: Dhan, Chaval.

Annuals. Culm erect, striate, smooth, nodes glabrous. Leaves linear, acuminate; sheaths smooth, striate; ligule 1 cm or more long; Panicle often 30 cm long, terminal, narrow. Spikelets laterally compressed, persistent, oblong, 6-8 mm long; pedicels scaberulous, 2-8 mm long. Glumes minute. Lowest 2 lemmas empty, lanceolate, 1-nerved. Uppermost lemma fertile, 5-nerved, awned. Palea cymbiform, acute. Lodicules 2, sub-quadrate. Cultivated rice plants of our areas are referable to var. *sativa* L. It is characterized by 1-seeded mature florets and smooth awns.

Ecology: Cultivated; somewhere as an escape.
Fl. & Fr.: Sept.-Dec.
Distribution: Throughout the plains and hills of India.
Specimens examined: Chordaha, 1146; Bukar, 5062.
Uses: Grains are used as an important food for human beings. The plant is also used as fodder; in making papers, cardboards, mats, etc. Rice-bran oil used for soaps, cosmetics and as an anticorrosion oil.

32. PANICUM L.

Key to the Species

1a. Lower glume cuspidate-acuminate. Spikelets gaping widely at anthesis ...... 4. *P. trypheron*
1b. Lower glume not cuspidate-acuminate. Spikelets not gaping:
  2a. Annual, not stoloniferous. Spikelets ovate-oblong or elliptic ...... 2. *P. psilopodium*
  2b. Perennial. Stems below creeping or stoloniferous. Spikelets lanceolate:
    3a. Leaf-blades broad, flat. Culms spongy, creeping and floating at base. Pedicels angular, scaberulous, with clavellate-truncate tips ...... 1. *P. paludosum*
    3b. Leaf-blades narrow, usually involute. Culms tough, stoloniferous at base. Pedicels glabrous, with small copular tips ...... 3. *P. repens*

1. Panicum paludosum Roxb., Fl. Indica 1: 310. 1820; Blatter & McCann, Bombay Grasses 162. 1935; Bor, Grasses Burma Ceyl. Ind. Pak. 329. 1960; Lazarides, Trop. Grasses S. E. Asia 130. 1980; Cope in Nasir & Ali, Fl. Pakistan 143: 169. 1982; SreeKumar & Nair, Fl. Kerala Grasses 268. 1991; Singh et al., Fl. Bihar 637. 2001; Paria & Chattopadhyay, Fl. Hazaribagh District 2: 1110. 2005; Kabir & Nair, Fl. Tamil Nadu Grasses 274. 2009. *Panicum proliferum* auct. non Lam. 1797; Hook. f., Fl. Brit. India 7: 50. 1896; Prain, Bengal Pl. 2: 1179. 1903 (Rep. ed. 2: 888. 1963); Haines, Bot. Bihar & Orissa pt. 5: 995. 1924.

Perennial. Culms erect or ascending, creeping and floating at base, rooting at lower nodes, spongy, 60-90 cm long. Leaf-blades linear or ensiform, flat, rounded or subcordate at base, scaberulous on margins, many-nerved, 15-30 cm long and 0.6-1.5 cm broad; sheaths lax, striate, glabrous; ligule reduced to a ring of hairs. Panicle at first contracted, then spreading, 10-25 cm long; branches fasciculate, below, trigonous, scabrous. Pedicels angular, scaberulous, clavellate-truncate at tips. Spikelets singly or in pairs, lanceolate, 3-4 mm long.

Lower glume clasping, orbicular or kidney-shaped, white, 0.6-0.8 mm long; upper ovate, strongly 7-9-nerved, 3-4 mm long. Lower floret: lemma ovate-lanceolate, 9-nerved, 3-4 mm long; palea small, or absent. Upper floret sessile; lemma narrowly oblong, dorsally convex, ivory or pale-yellow, with involute margins, 2.5-3.5 mm long. Palea narrowly oblong, with inturned margins and auricled base, 2.5-3.5 mm long. Anthers to 1.5 mm long, Caryopsis narrowly ellipsoid, ca 1 mm long.

Ecology: Common in marshy lowlands, rice fields, margins of tanks and others water courses.
Fl. & Fr.: Aug.-Dec.
Distribution: Throughout India. Pakistan, Nepal, Sri Lanka, through Myanmar to S.E. Asia and Australia.
Specimens examined: Muria, 1140; Mainukhar, 5073.
Uses: It provides a favourite fodder for elephants and buffaloes. Grains are used by hill tribes for making a cake-like preparation.

2. Panicum psilopodium Trin., Gram. Panic. 217. 1826; Hook. f., Fl. Brit. India 7: 46. 1896; Prain, Bengal Pl. 2: 1179. 1903; Haines, Bot. Bihar & Orissa pt. 5: 993. 1924; Bor, Grass. Burma Ceylon India & Pakistan 329. 1960; SreeKumar & Nair, Fl. Kerala Grasses 269. 1991; Singh et al., Fl. Bihar 638. 2001; Paria & Chattopadhyay, Fl. Hazaribagh District 2: 1111. 2005; Kabir & Nair, Fl. Tamil Nadu Grasses 275. 2009. *P. psilopodium* Trin. var. *coloratum* Hook. f., Fl. Brit. India 7: 47. 1896; Haines, Bot. Bihar & Orissa pt. 5: 993. 1924.

A tufted annual grass, 25-75 cm high. Leaves 10-30 x 0.4-1 cm, linear, acuminate, glabrous, or with few spreading hairs, base rounded; sheaths with spreading hairs; ligule hairy. Panicle 7-14 cm long, spreading; pedicels unequal, finely scabrid. Spikelets solitary or in pairs, 2.5-3 mm long, ovate-lanceolate, glabrous, flattened. Glume I suborbicular, 0.75-1 mm long, subacute or obtuse, 3-5 nerved. Glume II broadly ovate, 2.5 mm long, 9-11-nerved. Lower lemma empty, 2-25 mm long, ovate, acute, 9-nerved; palea elliptic. Upper lemma hermaphrodite, apiculate, shining; palea slightly shorter than glume.

Ecology: Common in rice-fields, wet lands, etc.
Fl. & Fr.: Aug.-Nov.

Distribution: Almost throughout India.
Specimens examined: Ahri, 1096; Silodhar, 5021.
Uses: It is used as cattle feed. Grains are used in the preparations of alcoholic beverages.

3. Panicum repens L., Sp. Pl. ed. 2. 87. 1762; Hook. f., Fl. Brit. India 7: 43. 1896; Prain, Bengal Pl. 2: 1179. 1903 (Rep. ed., 2: 888, 1963); Haines, For. Fl. Chotanagpur 562. 1910 & Bot. Bihar & Orissa pt. 5: 994. 1924; Bor, Grasses Burma Ceyl. Ind. Pak. 330. 1960; Lazarides, Trop. Grasses S. E. Asia 130. 1980; SreeKumar & Nair, Fl. Kerala Grasses 271. 1991; Singh et al., Fl. Bihar 638. 2001; Paria & Chattopadhyay, Fl. Hazaribagh District 2: 1112. 2005; Kabir & Nair, Fl. Tamil Nadu Grasses 277. 2009.

Perennials, rhizomatous. Culms 40-100 cm tall; nodes glabrous. Leaves linear-lanceolate, glabrous; sheaths ciliate at throat; ligules hairy. Inflorescence an open panicle; branches scabrid. Spikelets elliptic-lanceolate, 2.5-3 mm long, glabrous. Lower glume ovate to orbicular, 0.5-1 mm long; upper glume elliptic-lanceolate, 2.5-3 mm long, acute, membranous, glabrous, 5-9 nerved. Lower floret male. Upper floret bisexual. Lemma elliptic-oblong, 2.5-2.7 mm long.

Ecology: Common in aquatic, semi-aquatic habitat and cultivated field.
Fl. & Fr.: Aug.-Sept.
Distribution: Throughout India.
Specimens examined: Sikda, 1132; Mainukhar, 5016.
Uses: It is a good fodder grass and is also used for turfs and lawns.

4. Panicum trypheron Schult., Syst. Veg. 2, Mant. 244. 1824; Hook. f., Fl. Brit. India 7: 47. 1896; Haines, Bot. Bihar & Orissa pt. 5: 995. 1924 (Repr. ed., 3: 1041. 1961); Bor, Grass. Burm. Ceyl. Ind. Pak. 331. 1960; SreeKumar & Nair, Fl. Kerala Grasses 273. 1991; Singh et al., Fl. Bihar 638. 2001; Paria & Chattopadhyay, Fl. Hazaribagh District 2: 1113. 2005.

Annual grass. Glumes 30-80 cm high, tufted. Leaves 8-23 x 0.35-0.55 cm, linear, acuminate, thin, sparsely hairy; sheaths long, glabrous or hairy; ligule short, hairy. Panicle often large and 10-25 cm long; pedicels 3-4 mm long. Spikelets 2.5-3 mm long, ovoid, acuminate, green with purple tinge. Glume I 2 mm long, broadly ovate, cuspidate, 5-nerved. Glume II 3-3.25 mm long, broadly ovate, usually 9-nerved. Lower lemma 3 mm long; palea elliptic-oblong, white. Upper lemma 2-2.5 mm, elliptic oblong; palea elliptic.

Ecology: Frequent in pasture lands and open places of the forests.
Fl. & Fr.: Aug.-Dec.
Distribution: Throughout the plains and hills of India.
Specimens examined: Morainia, 5009; Kathodumar, 5078.
Uses: Plant is grazed by cattle. Grains are used in making breads in the time of scarcity.

33. PASPALIDIUM Stapf

Key to the Species

1a.	Annuals; spikes shorter than the internodes; upper lemma granular	...... *P. flavidum*
1b.	Perennials; spikes longer than the internodes upper lemma not granular	...... *P. geminatum*

1. Paspalidium flavidum (Retz.) A. Camus in Lecomte, Fl. Gen., Indo-China 7: 419. 1922; Haines, Bot. Bihar & Orissa pt. 5: 1001; Bor, Common Grasses of the United Provinces 172. 1940; Bor, Grasses Burm. Ceyl. Ind. Pak. 333. 1960; Lazarides. Trop. Grasses S.E. Asia 131. 1980; Cope in Nasir & Ali, Fl. Pakistan 143: 190. 1982; SreeKumar & Nair, Fl. Kerala Grasses 278. 1991; Singh et al., Fl. Bihar 639. 2001; Paria and Chattopadhyay, Fl. Hazaribagh District 2: 1115. 2005; Kabir & Nair, Fl. Tamil Nadu Grasses 281. 2009. *Panicum flavidum* Retz. Obs. Bot. 4: 15. 1786; Hook f., Fl. Brit. India 7: 28. 1896; Prain, Bengal Pl. 2: 1176. 1903 (Rep. ed. 2: 855. 1963).

Local name: Chapri.

Glabrous annual or perennial, 30-100 cm in ht. Culms slender, compressed, decumbent-ascending. Leaves 7-13 x 0.3-0.7 cm, linear, acuminate, glabrous; ligule a ridge of hairs. Panicle with 5-9 spikes, often 25-30 cm long; spikes 1.2-2.5 cm long, recurved, alternate; rachis flattened, with two rows of spikelets. Spikelets subsessile, 8-20, yellowish, 2-3 mm long, ovate, glabrous. Gl. I suborbicular, 1-1.25 x 1.25-1.5 mm, concave, 3-nerved. Gl. II suborbicular,

2 mm long, 7-nerved. Lower lemma male or neuter, 2.5-3 mm; palea 2.25-2.5 mm long, hyaline. Upper lemma bisexual, 2.25-2.5 mm broadly ovate, mucronulate; palea 2 mm long (Illus. Plate-XX; Fig.-22 and Plate-XXXX; Fig.-57A and B).

Ecology: Frequent in moist places and roadsides.
Fl. & Fr.: Aug.-Nov.
Distribution: Throughout the moisture parts of plains and hills of India. Tropical Asia.
Specimens examined: Bukar, 1009; Asnachuan, 5074.
Uses: A good fodder grass.

2. Paspalidium geminatum (Forssk.) Stapf in Prain, Fl. Trop. Afr. 9: 583. 1920; Haines, Bot. Bihar Orissa pt. 5: 1002. 1924 (Repr. ed., 3: 1048. 1961); Bor, Grass Gurm. Ceyl. Ind. Pak. 33. 1960; SreeKumar & Nair, Fl. Kerala Grasses 279. 1991; Singh et al., Fl. Bihar 639. 2001; Kabir & Nair, Fl. Tamil Nadu Grasses 282. 2009. *Panicum geminatum* Forssk., Fl. Aegypt.-Arab. 18. 1775. *P. paspaloides* Pers., Sys. Pl. 1: 81. 1805; Hook. f., Fl. Brit. India 7: 30. 1896.

Perennials, stoloniferous. Culms 40-100 cm tall, subcompressed, glabrous. Leaves glabrous; sheaths glabrous, compressed; ligules hairy. Inflorescence of several racemes appressed to rachis, alternate, longer than internodes. Spikelets solitary, secund, ovate-acute, 2.5-3 mm long, glabrous. Lower glume hyaline, 0.5-0.7 mm long, nerveless; upper glume 1.8-2 mm long, convex, membranous, 5-nerved. Lower floret empty. Upper floret bisexual. Lemma elliptic, 2-2.2 mm long, apiculate, coriaceous, 5-nerved. Anthers 3, yellow, 1-1.5 mm long. Caryopsis broadly ellipsoid, ca 2 mm long.

Ecology: Frequent along the water-lodged areas, shallow watered ditches, etc.
Fl. & Fr.: July-Dec.
Distribution: Throughout India.
Specimens examined: Sanjha, 1139; Pathalgara, 5097.

34. PASPALUM L.

Key to the Species

1a.	Rachis narrower than distichous spikelets	...... 1. *P. distichum*
1b.	Rachis as wide as the spikelets	...... 2. *P. scrobiculatum*

1. Paspalum distichum L., Syst. Nat. ed. 10, 2: 855. 1759; Hook. f., Fl. Brit. India 7: 12. 1896; Prain, Bengal Pl. 2: 1183. 1903; Bor, Grass. Burma Ceylon India & Pakistan 338. 1960; SreeKumar & Nair, Fl. Kerala Grasses 286. 1991; Paria & Chattopadhyay, Fl. Hazaribagh District 2: 1116. 2005; Kabir & Nair, Fl. Tamil Nadu Grasses 287. 2009.

Perennial, with subcompressed culms and slender rhizome, stoloniferous; stolons slender, subcompressed, to 1m long; nodes dark, hairy. Leaf-blades flat, with rounded ciliate base, acuminate at apex, 3-12 cm long and 2-6 mm wide; sheaths loose, keeled, pilose on margins; ligule membranous, ca 3 mm long. Racemes 2, rarely 4, erect or reflexed, incurved, 1.5-7 cm long; rachis usually pedunculate in one, sometimes in both racemes, with long hairs in axils, triquetrous, scaberulous on margin. Spikelets solitary, rarely in pairs, 2.5-3.5 mm long, elliptic; lower glume occasionally developed. Upper glume and sterile lemma equal, 3-5-nerved, the former appressed-pubescent. Caryopsis elliptic, 2.5-2.8 mm long.

Ecology: Common along the banks of backwaters and streams, as a weed in paddy fields.
Fl. & Fr.: Aug.-Nov.
Distribution: Throughout India.
Specimens examined: Kabilas, 1061; Silodhar, 5008.
Uses: Plant forms a valuable pasture grass and soil binder.

2. Paspalum scrobiculatum L., Mant. Pl. 1: 29. 1767; Hook. f., Fl. Brit. India 7: 10. 1896; Prain, Bengal Pl. 2: 1182. 1903; Haines, For. Fl. Chota Nagpur 560. 1910 & Bot. Bihar & Orissa pt. 5: 1000. 1924; Bor, Common Grasses of the United Provinces 174. 1940; Bor, Grass. Burma Ceylon India & Pakistan 340. 1960; Cope in Nasir & Ali, Fl W. Pakistan 143: 213. 1982; SreeKumar & Nair, Fl. Kerala Grasses 287. 1991; Singh et al., Fl. Bihar 640. 2001; Paria & Chattopadhyay, Fl. Hazaribagh District 2: 1117. 2005; Kabir

& Nair, Fl. Tamil Nadu Grasses 288. 2009. *P. orbiculare* Forst., Insul. Austr. Prodr. 7. 1786; Bor, Grass. Burma Ceylon India & Pakistan 340: 1960. *P. commersonii* Lam., Tab. Ency. Meth. Bot. 1: 175. 1791; Bor, Grass. Burma Ceylon India & Pakistan 340: 1960. *P. cartilageneum* Presl, Rel. Haenk. 1: 216. 1830, Bor, Grass Ceylon India & Pakistan 335. 1960.

Annual, 60-125 cm high. Culms tufted, rooting below. Leaves 10-45 x 0.2-0.8 cm, finely acuminate, base keeled; sheath 10-20 cm long; loose, mouth fairy; ligules short, membranous. Inflorescence 4-7 cm long, of 2-6 distant, false recemes; pedicel very short or sessile. Spikelets biseriate, broadly elliptic or suborbicular, imbricate. Gl. I absent. Gl. II broadly elliptic with rounded apex, 2-2.5 mm long, 5-nerved. Lower lemma broadly elliptic, 2 mm long 5 to 7-nerved; palea 2 mm long, membranous, sometimes wanting. Upper lemma suborbicular; palea 1.75-2 mm long. Caryopsis biconvex (Illus. Plate-XXI; Fig.-23).

Ecology: Common under shady trees, beside nalas, in the ravines and in paddy fields.
Fl. & Fr.: Aug.-Nov.
Distribution: Throughout India and all the warm countries.
Specimens examined: Garmorwa, 1006; Duragara, 5047.
Uses: It is used as fodder for cattle.

35. PENNISETUM Rich.

Pennisetum pedicellatum Trin. in Mem. Acad. Sci. Petersb. Ser. 6, 3: 184. 1834; Hook f., Fl. Brit. India 7: 86. 1896; Haines, Bot. Bihar & Orissa pt.5: 986. 1924 (Repr. ed., 3: 1032. 1961); Bor, Grass. Burm, Ceyl. Ind. Pak. 346. 1960; SreeKumar & Nair, Fl. Kerala Grasses 290. 1991; Singh et al., Fl. Bihar 641. 2001; Paria & Chattopadhyay, Fl. Hazaribagh District 2: 1118. 2005; Kabir & Nair, Fl. Tamil Nadu Grasses 294. 2009.

Annual, often gregarious, 30-90 cm high. Leaves 10-30 cm long, 5-13 mm broad, linear, acuminate, flat, long hairy along the margins near the base; sheaths glabrous; ligule a shortly ciliate membrane. Inflorescence 8-13 x 1.5-2 cm, purple, spiciform, dense; involucres sessile; outer bristles few to many; inner bristles numerous. Spikelets solitary or geminate, oblong-lanceolate, 4-5 mm long; pedicels woolly. Glume I oblong-lanceolate, 2.75-3 mm long, acute, hyaline, silky-hairy. Glume II ovate-oblong, 5-nerved. Lower lemma male, 3.25 mm long, 3-toothed; palea 2.5 mm long, hyaline. Upper lemma female, 2 mm long, lanceolate, shining; palea lanceolate. Caryopsis narrowly oblong, 1.5 mm (Plate-XXXX; Fig.-58).

Ecology: Common along roadsides, in forests, etc.
Fl. & Fr.: Sept.-Dec.
Distribution: Bihar, Jharkhand, West Bengal, etc.
Specimens examined: Sikda, 1108; Chordaha, 5089.
Uses: A good fodder grass.

36. PEROTIS Ait.

Perotis indica (L.) O. Kuntze, Rev. Gen. Pl. 2: 787. 1891; Bor, Grass. Burma Ceylon India & Pakistan 611. 1960; SreeKumar & Nair, Fl. Kerala Grasses 434. 1991; Singh et al., Fl. Bihar 642. 2001; Paria & Chattopadhyay, Fl. Hazaribagh District 2: 1119. 2005; Kabir & Nair, Fl. Tamil Nadu Grasses 205. 2009. *Anthoxanthum indicum* L., Sp. Pl. 28. 1753. *Perotis latifolia* Ait., Hort. Kew. 1: 85. 1789; Roxb., Fl. Ind. 1: 239. 1820; Hook. f., Fl. Brit. India 7: 98. 1896; Prain, Bengal Pl. 2: 1186. 1903; Haines, Bot. Bihar & Orissa pt. 5: 978. 1924.

Annuals. Culms 40-50 cm tall, erect. Leaves narrowly ovate, 1-3.5 cm long, cordate at base, scabrid on margins; sheaths terete, glabrous; ligules membranous, truncate. Inflorescence a terminal spike. Spikelets 1.5-2.5 mm long, 2-awned, hispid. Lower glume lanceolate, ca 1.9 mm long, 1-nerved awn ca 1 cm long; upper glume similar to lower, with ca 7.5 mm long awn. Lemma hyaline, lanceolate, ca 0.6 mm long, 1-nerved. Palea nerveless. Anthers yellow, ca 0.4 mm long. Caryopsis brown, terete, ca 1.4 mm long.

Ecology: Occurs in overgrazed sandy pasture lands.
Fl. & Fr.: Oct.-Nov.
Distribution: Almost throughout India, Nepal, Sri Lanka, Myanmar, and Malesia to South - East Asia.
Specimens examined: Garmorwa, 1086; Pathalgara, 5087.
Uses: It forms a good fodder grass relished by stock at all stages.

37. PHALARIS L.

Phalaris minor Retz. Observ. Bot. 3: 8. 1783; Hook. f., Fl. Brit. India 7: 221. 1896; Bor, Grass. Burm. Ceyl. Ind. Pak. 616. 1960; Singh et al., Fl. Bihar 642. 2001; Paria & Chattopadhyay, Fl. Hazaribagh District 2: 1120. 2005; Kabir & Nair, Fl. Tamil Nadu Grasses 84-85. 2009.

Annual. Culms erect or ascending, 30-90 cm long, stout or slender, leafy. Leaf-blades linear, acuminate, glabrous, 15-25 cm long and 1-1.5 cm broad; sheath smooth; ligule oblong, scabrous. Panicle dense, ovoid or cylindric, green, 6-9 cm long and 1.5-2 cm in diam. Spikelets ellipsoid, shortly pedicellate, shining, 5-6 mm long and 3-3.5 mm broad. Glumes acuminate; wings irregularly crenate or serrulate. Fertile lemma ovate, silky, with one bristle-like imperfect glume at base. Palea lanceolate, ciliate at top. Anthers pale yellow. Caryopsis ovoid, apiculate, 1.8-2 mm long (Plate-XXXXI; Fig.-59).

Ecology: Mostly occur in fallow fields.
Fl. & Fr.: Feb.-Apr.
Distrubtion: Kashmir, plains of western, central and eastern India.
Specimens examined: Mainukhar, 1036; Ahri, 5002.
Uses: Plant is consumed by dairy cattle. Grains are used as a bird feed.

38. POLYPOGON Desf.

Polypogon monspeliensis (L.) Desf., Fl. Atl. 1: 67. 1798; Hook. f., Fl. Brit. India 7: 245. 1896; Haines, Bot. Bihar & Orissa pt. 5: 976. 1924 (Repr. ed., 3: 1021. 1961); Bor, Grass. Burm. Ceyl. Ind. Pak. 403. 1960; Singh et al., Fl. Bihar 644. 2001; Paria & Chattopadhyay, Fl. Hazaribagh District 2: 1123. 2005. *Alopecurus monspeliensis* L. Sp. Pl. 61. 1753.

Culms tufted, stout or slender, from a geniculate base, 15-60 cm long. Leaves linear, flat, scabrid; sheaths striate, inflated, scaberulous; ligule lanceolate, ciliolate, to 6 mm long. Panicle cylindric, 3.5-12 cm long, light green or straw-coloured; branches scaberulous; pedicels disarticulating near base. Spikelets 2-2.5 mm long. Glumes subequal, linearly oblanceolate-oblong, shortly 2-lobed or emarginate, scaberulous or ciliolate, with to 8 mm long awn. Lemma ovate, 1.2-1.3 mm long, minutely 2-4-mucronulate, obscurely nerved;

awn as long as lemma or shorter or absent. Palea 2-toothed or 2-mucronulate. Caryopsis subterete, grooved, 1.3-1.5 mm long (Plate-XXXXI; Fig.-60).

Ecology: Common in fallow fields, on sandy dried river-beds, sandy grounds, etc.
Fl. & Fr.: Feb.-April.
Distribution: India: Jammu and Kashmir, Himachal Pradesh, Punjab, Delhi, Uttar Pradesh, Rajasthan, Madhya Pradesh, Jharkhand, Bihar, West Bengal, Meghalaya, Andhra Pradesh, Maharashtra, Karnataka, Tamil Nadu, Europe and temperate parts of Asia and Africa.
Specimens examined: Kathodumar, 1035; Sanjha, 5018.
Uses: Plant affords rich feeding for grazing animals. Sometimes it is cultivated in gardens for tis beautiful panicle.

39. **ROTTBOELLIA** L. f., nom. cons.

Rottboellia cochinchinensis (Lour.) W. D. Clayton in Kew Bull. 35: 817. 1981; SreeKumar & Nair, Fl. Kerala Grasses 183. 1991; Singh et al., Fl. Bihar 646. 2001; Paria & Chattopadhyay, Fl. Hazaribagh District 2: 1125. 2005; Kabir & Nair, Fl. Tamil Nadu Grasses 461. 2009. *Stegosia cochinchinensis* Lour., Fl. Cochinch. 1: 51. 1790. *Rottboellia exaltata* L. f., Suppl. Pl. 114. 1781, non (L.) L. f., 1779; Hook. f., Fl. Brit. India 7: 156. 1896; Haines, Bot. Bihar & Orissa pt. 6: 1059. 1924; Bor, Common Grasses of the United Provinces 183. 1940; Bor, Grass. Burm. Ceyl. Ind. Pak. 206. 1960.

Local name: Bhursali, Barsali.

Annuals. Culms to 3m tall, branched, terete with a channel facing the branches. Leaf-blades linear-lanceolate, tapering to a long fine point, somewhat flaccid, smooth below, scabrid above and along margins, 30-45 cm long and 1.2-2.5 cm wide; sheaths terete, tight or widened upwards, hirsute, striate; ligules brown, stout, rounded, ciliolate. Racemes 7.5-15 cm long, stiff, pale yellowish–green; joints 5-8 mm long, rounded on back, concave on inner face, smooth, disarticulating with an orbicular scar; pedicels shorter than joints, rounded on back, concave on inner face. Sessile spikelet ovate-lanceolate, glabrous, 6-8 mm long; callus broad, smooth. Lower glume entire or minutely 2-toothed, usually narrowly winged near emarginate apex, with about 9 faintly visible intracarinal nerves on inner surface; upper glume very broad in profile, about 11-nerved. Lower floret: lemma oblong-lanceolate, glabrous, acute, 5-6 mm long; palea similar to lemma, rigid, 2-nerved. Upper floret: lemma slightly shorter than that of lower floret, boat- shaped, obliquely ovate in profile; palea narrowly oblong, nerveless, nearly as long as lemma. Caryopsis 2.5-3 x 1.5-2 mm (Illus. Plate-XXII; Fig.-24).

Ecology: Common in grassy open places of forests.
Fl. & Fr.: Sept.-Dec.
Distribution: India and throughout Old World tropics and introduced to the Caribbean.
Specimens examined: Garmorwa, 5010; Sikda, 6034.
Uses: A good fodder grass; also used in making mats.

40. SACCHARUM L.

Key to the Species

1a. Rachis very fragile; peduncle silvery white hairy below the panicle ...... 3. *S. spontaneum*

1b. Rachis less fragile; peduncle not hairy below the panicle:

- 2a. Leaf sheath hairy at base, villous on margins; spikelets 4-6 mm long; lower glume of sessile spikelet with long hairs ...... 1. *S. bengalense*
- 2b. Leaf sheath pubescent towards throat, otherwise glabrous; spikelets 3.5-4 mm long; lower glume of sessile spikelet glabrous ...... 2. *S. officinarum*

1. Saccharum bengalense Retz. Obs. Bot. 5: 16. 1789: GBCIP 211; Singh et al., Fl. Bihar 646. 2001; Paria & Chattopadhyay, Fl. Hazaribagh District 2: 1126. 2005. *S. munja* Roxb. Fl. Ind. 1: 250. 1820; Bor, Common Grasses of the United Provinces 187. 1940; *S. arundinaceum* Hook. f., Fl. Brit. Ind. 7: 119. 1896 pro parte (non Retz. 1789); BP 2: 895; Haines, Bot. Bihar & Orissa pt. 5: 1012. 1924; Bor, Common Grasses of the United Provinces 189. 1940. *Erianthus munja* (Roxb.) Jesweit, Arch. Suikerind. Ned.-Ind. 399. 1925.

Local name: Bada Kasi.

Perennial grass, 3.5-5.5 mm tall. Culms erect from a stout root-stock. Leaves 1.5-2 mm long near the base, 1.6-2.5 cm broad, tapering to a slender point at the apex, glaucous, margins, spinulose; sheath bearded about the mouth; ligule hairy. Panicle terminal, 30-60 cm long, covered by woolly hairs; pedicel 3.5 mm. Sessile spikelets: gl. I 4.75 mm long, lanceolate, acuminate, 3-nerved, villous on the back. Gl. II 4.5-4.75 mm long, cymbiform, acuminate, 2-keeled. Lower lemma empty, 3.75 mm long, epaleate, 1-nerved. Upper lemma 3.75 mm long, oblong-lanceolate, aristulate with 1 mm long awn; palea hyaline. Caryopsis 2 mm, linear-oblong. Pedicelled spikelets similar to sessile spikelets (Plate-XXXXII; Fig.-61).

Ecology: Usually gregarious on the sandy river beds or on sandy alluvial soil in the neighbourhood of streams, along roadsides, etc.
Fl. & Fr.: Oct.-Dec.
Distribution: North, North-Western and Central India, Bihar, Jharkhand, Orissa and West Bengal.
Specimens examined: Hathia Baba, 1110; Danua, 1099.
Uses: The fibre of the leaf - sheath is used for making mats and ropes.

2. Saccharum officinarum L., Sp. Pl. 54. 1753; Hook. f., Fl. Brit. India 7: 118. 1896; Haines, Bot. Bihar & Orissa pt. 5: 1012. 1924; Bor, Grass. Burm. Ceyl. Ind. Pak. 212. 1960; SreeKumar & Nair, Fl. Kerala Grasses 185. 1991; Singh et al., Fl. Bihar 506. 2001.

Local name: Ketari

Perennials. Culms 2-6 m tall, erect, glabrous, waxy below the nodes. Leaves linear, very long, acute at apex, scabrous on margins; sheaths pubescent towards throat; ligules ciliate. Inflorescence a large, pyramidal panicle; nodes bearded; rachis breaking up at maturity; pedicels glabrous. Spikelets 3.5-4 mm long, covered with white silky hairs; callus with long hairs. Lower glume equalling spikelet, acute, margins inserted; upper glume similar to lower. Lower floret empty. Upper floret bisexual. Lemma awnless, reduced. Anthers 3, purplish, ca 2 mm long. Caryopsis purple, cylindrical, ca 1.5 mm long.

Ecology: Cultivated in lands and along the rivers/streams of the area.

Fl. & Fr.: Feb.-April.
Distribution: Almost throughout India.
Specimens examined: Pathalgarwa, 7001; Chordaha, 7008.
Uses: 'Gud' and 'chini' are made from this; also juice of this is used by the people. Green leaves are eaten by cattles.

3. Saccharum spontaneum L., Mant. 183. 1771; Hook. f., Fl. Brit. India 7: 118. 1896; Prain, Bengal Pl. 2: 1188. 1903 (Rep. ed., 2: 895. 1963); Haines, Bot. Bihar & Orissa pt. 5: 1011. 1924; Punj, in Indian J. Agric. Sc. 3: 1013. 1933; Bor, Common Grasses of the United Provinces 185. 1940; Mukerjee in Bull. Bot. Soc. Bengal 8: 145. 1949; Bor, Grasses Burm. Ceyl. Ind. Pak. 214. 1960; Cope in Nasir & Ali, Fl. Pakistan 143: 263. 1982; SreeKumar & Nair, Fl. Kerala Grasses 185. 1991; Singh et al., Fl. Bihar 648. 2001; Paria & Chattopadhyay, Fl. Hazaribagh District 2: 1128. 2005; Kabir & Nair, Fl. Tamil Nadu Grasses 465. 2009.

Local name: Kasi.

A tall, erect, perennial grass, 1.5-4.5 m high. Culms solid, smooth, glaucous, silky below the panicle. Leaves 30-70 x 0.3-0.7 cm, linear, finely acuminate, glabrous, margins finely spinulose; sheath with hairy mouth; ligule membranous. Panicle 20-60 cm long, lanceolate, white silky hairy; rachis capillary, fragile. Spikelets lanceolate, 1-flowered, bisexual, 4-5 mm long; callus short, silky hairs of callus many times longer than the spikelets. Gl. I lanceolate, 4-5 mm long. Gl II 4-4.5 mm, broadly lanceolate, l-nerved. Lower lemma, 3.5 mm long, ovate-lanceolate, ciliate. Upper lemma minute, linear, hyaline; palea minute, ciliate. Sessile and pedicellate spikelets similar (Plate-XXXXII; Fig.-62).

Ecology: Grows gregariously along sandy river banks, on lands and along roadside. The leaves are used for thatching.
Fl. & Fr.: Aug.-Nov.
Distribution: Throughout the warmer parts of India and warmer regions of Old World.
Specimens examined: Dhoria, 1087; Chordaha, 1090; Kabilas, 1103.
Uses: Young part of the plant is used as fodder only in times of scarcity. Plant yields a good paper-pulp. Culms are used for thatching.

41. SACCIOLEPIS Nash

Key to the Species

1a.	Spikelets obliquely lanceolate, acute at tips	...... 1. *S. interrupta*
1b.	Spikelets elliptic, blunt at tips	...... 2. *S. myosuroides*

1. Sacciolepis interrupta (Willd.) Stapf in Prain, Fl. Trop Afr. 9: 757. 1920; Haines, Bot. Bihar Orissa pt. 5: 991. 1924; Bor, Grass. Burm. Ceyl. Ind. Pak. 358. 1960; SreeKumar & Nair, Fl. Kerala Grasses 299. 1991; Singh et al., Fl. Bihar 648. 2001; Kabir & Nair, Fl. Tamil Nadu Grasses 303. 2009. *Panicum interruptum* Willd., Sp. Pl. 1: 341. 1798; Hook. f., Fl. Brit. India 7: 40. 1896.

A glabrous perennial grass. Culms 1-1.5 m long, stout and spongy below; lower nodes bearing stout roots. Leaves 15-30x0.6-1.2 cm, linear, finely acuminate, flat; sheaths 5-12 cm long, with smooth margins; ligule thin, short. Panicle 15-30 x 0.6-0.8 cm, spiciform, interrupted below. Spikelets green, 4-5 mm long, in interrupted crowded fascicles, ovate-lanceolate, acute. Gl. I broadly ovate, 1.5 mm long, hyaline, 3-5-nerved. Gl. II ovate, 4.5 mm long. Lower lemma 4.5 mm long, ovate; palea 3 mm long, hyaline, male or barren. Upper lemma 3 mm long, ovate-oblong, white; palea hyaline. Caryopsis obovoid (Plate-XXXXIII; Fig.-63).

Ecology: Common in swampy places and stagnant water of the forests and as a weed in paddy fields.
Fl. & Fr.: Sept.-Nov.
Distribution: Throughout India.
Specimens examined: Duragara, 1142; Pathalgara, 6014.

2. Sacciolepis myosuroides (R. Br.) A. Camus in Lecomte, Fl. Indo-China 7: 460. 1922; Haines, Bot. Bihar & Orissa pt. 5: 990. 1924; Bor, Grass. Burm. Ceyl. Ind. Pak. 358. 1960; SreeKumar & Nair, Fl. Kerala Grasses 300. 1991; Singh et al., Fl. Bihar 648. 2001; Kabir & Nair, Fl. Tamil Nadu Grasses 304. 2009. *Panicum myosuroides* R. Br., Prodr. 189. 1810.

Annuals. A slender grass. Culms 40-60 cm in ht., glabrous. Leaves 12-24 x0.25-0.55 cm, linear, acuminate, margin smooth; sheath striate, glabrous; ligule absent. Inflorescence a narrow spikelike panicle 8-15 cm long, dense, cylindric; rachis slender; pedicel very short, dilated at the tip. Spikelets green or purplish, densely crowded in small fascicles, obtuse, glabrous, 1.5 mm long. Gl. I ovate, 3-nerved, 0.5-1 mm long. Gl II ovate-oblong, obtuse, 5-9-nerved. Lower lemma empty, 1.25 mm long, ovate, 9-nerved; palea hyaline. Upper lemma bisexual, 0.5 mn long; palea as long as lemma.

Ecology: Common in shady and moist places, along ditches, water-lodged fields and as a weed in cultivated land.
Fl. & Fr.: Aug.-Dec.
Distribution: Throughout India.
Specimens examined: Silodhar, 1143; Murtiakalan, 6061.
Uses: As a fodder.

42. **SEHIMA** Forssk.

Sehima nervosum (Rottb.) Stapf in Prain, Fl. Trop. Afr. 9: 36. 1917; Haines, Bot. Bihar & Orissa pt. 5: 1023. 1924; Bor, Common Grasses of the United Provinces 196. 1940; Bor, Grass. Burm. Ceyl. Ind. Pak. 218. 1960; SreeKumar & Nair, Fl. Kerala Grasses 189. 1991; Singh et al., Fl. Bihar 650. 2001; Kabir & Nair, Fl. Tamil Nadu Grasses 468. 2009. *Andropogon nervosus* Rottl. Apud Willd. In Verh. Ges. Naturf. Freund. Berlin, Neve Schr 4: 218. 1803. *Ischaemum laxum* R. Br., Prodr. 205. 1810; Hook f., Fl. Brit. India 7: 136. 1896.

Perennials. Culms up to 90 cm tall, erect, tufted. Leaves linear, filiform at tips, rigid, scabrous; sheaths glabrous, striate; ligules hairy. Inflorescence a solitary raceme. Sessile spikelets linear-lanceolate, 6-8 mm long; callus shortly bearded. Lower glume 6-7 mm long, 2-dentate, deeply grooved on back; upper glume cymbiform, 5.5-6.5 mm long; arista ca 1.5 mm long. Lower floret male. Upper floret bisexual. Lemma deeply clefted into two lobes; awn ca 2.5 cm long. Anthers 3, yellow, 3-4 mm long. Pedicellate spikelets equal to the sessile ones. Caryopsis narrowly cylindrical, ca 2 mm long (Illus. Plate-XXIII; Fig.-25).

Ecology: Common in Scrub jungles and in rocky areas.
Fl. & Fr.: Aug.-Dec.

Distribution: Throughout India.
Specimens examined: Pathalgara, 1149; Chordaha, 6029.
Uses: Excellent fodder.

43. SETARIA Beauv.

Key to the Species

1a. Leaf blades folded or flat; inflorescence a cylindrical, dense or lobed spiciform panicle:
  2a. Bristles retrorsely barbed ...... 4. *S. verticillata*
  2b. Bristles antrorsely barbed:
    3a. Inflorescence a cylindrical false spike; spikelets 2-3 mm long:
      4a. Spikelets ca 3 mm long; upper lemma coarsely rugose, cymbiform, slightly keeled upwards, dorsally strongly curved in profile ...... 1. *S. glauca*
      4b. Spikelets 2-2.5 mm long; upper lemma finely rugose, narrow, not keeled, dorsally gently curved ...... 3. *S. pumila*
    3b. Inflorescence a narrow panicle; Spikelets 1.8-2 mm long ...... 2. *S. intermedia*

1. Setaria glauca (L.) P. Beauv., Ess. Agrost. 51, 169 & 178. 1812; Hook. f., Fl. Brit. India 7: 78 1896; Prain, Bengal Pl. 2: 1170. 1903; Haines, For. Fl. Chota Nagpur 560. 1910 & Bot. Bihar & Orissa pt. 5: 988. 1924; Bor, Common Grasses of the United Provinces 198. 1940; Bor, Grass. Burma Ceylon India & Pakistan 360. 1960; Paria & Chattopadhyay, Fl. Hazaribagh District 2: 1129. 2005. *Panicum glaucum* L., Sp. Pl. 56. 1753.

Local name: Bandra.

Annual grass, 30-60 cm high. Culms simple or branched. Leaves 10-30 x 0.7-1 cm, linear, ending in a long filiform tip, glabrous or sparsely hairy; sheaths smooth, glabrous; lilgule reduced to a ridge of hairs. Inflorescence brownish-yellow, 3-10 x 0.7-1 cm, spiciform; rachis puberulous; bristles 6-8 to each involucre; pedicel with a discoid tip. Spikelets closely set along the rachis, 3 mm. Glume I ovate, 1.5 mm long, 3-nerved. Glume II broadly ovate, 2-2.5 mm, 5-nerved. Lower lemma male or barren, broadly ovate, or oblong, 5-nerved; palea 2-2.25 mm or less. Upper lemma cymbiform transversely rugose; palea with stout keels (Illus. Plate-XXIV; Fig.-26 and Plate-XXXXIII; Fig.-64A and B).

Ecology: Frequent in grassy lands and sometimes among other crops.
Fl. & Fr.: Aug.-Oct.
Distribution: Throughout India.
Specimens examined: Kenduadih, 1020; Sikda, 6005.
Uses: A good fodder grass.

2. Setaria intermedia Roem. & Schult. Syst. Veg. 2: 489. 1817; Hook. f., Fl. Brit. India 7: 79. 1896; Prain, Bengal Pl. 2: 1170. 1903 (Rep. ed. 2: 881. 1983); Haines, Bot. Bihar & Orissa pt. 5: 989. 1924; Blatter & McCann, Bombay Grasses 174. 1935; Cope in Nasir & Ali. Fl. Pakistan 143: 183. 1982; SreeKumar & Nair, Fl. Kerala Grasses 306. 1991; Singh et al., Fl. Bihar 651. 2001; Paria & Chattopadhyay, Fl. Hazaribagh District 2: 1130. 2005; Kabir & Nair, Fl. Tamil Nadu Grasses 308. 2009. *Panicum temontosum* Roxb. Fl. Indica 1: 303. 1820. *Setaria tomentosa* (Roxb.) Kunth, Rev. Gram. 1: 47. 1929; Bor, Grasses Burm. Ceyl. Ind. Pak. 365. 1960. *P. intermedium* (Roem. & Schult.) Roth. Nov. Pl. Sp. 47. 1821.

Annual. Culms tufted, geniculate, minutely scabrid, to 1 m high. Leaf-blades linear, rounded at base, flat, flaccid, hairy, rough on margins; sheaths thin, ciliate along margins, usually bearded at mouth, somewhat keeled and compressed; ligule reduced to a long-ciliate rim. Panicle straight or flexous, ± loose, narrow, lobed in lower part, tapering upwards, 2.5-10 cm long; lower branches to 1.5 cm long, upper reduced to subsessile clusters of usually 2 or solitary spikelets supported by antrorsely scabrid bristles of 3-3.5 mm long; pedicels with discoid tips. Spikelets ovate-ellipitic, 2-2.5 mm long. Lower glume ovate, half as long as spikelets; upper elliptic, concave, 2/3 length of upper floret. Lower floret barren; lemma elliptic-oblong, dorsally compressed; palea elliptic, shorter than lemma, marginately keeled. Upper floret; lemma boat-shaped, transversely rugose; palea slightly rugose. Caryopsis ovoid, plano-convex, 1.2-1.5 mm long (Plate-XXXXIV; Fig.-65A and B).

Ecology: Common in damp shady places, gardens, etc.
Fl. & Fr.: July-Oct.
Distribution: India: Himalaya, Punjab to West Bengal, Assam, Bihar, Jharkhand, Orissa, Central and Southern India. Myanmar, Sri Lanka, extending to Malesia and Polynesia.
Specimens examined: Chordaha, 1055; Morainia, 6010.
Use: Plant is used as cattle feed.

3. Setaria pumila (Poir.) Roem. & Schult., Syst. Veg. 2: 891. 1817; Clayton in Tutin et al., Fl. Eur. 5: 263. 1980; SreeKumar & Nair, Fl. Kerala Grasses 309. 1991; Singh et al., Fl. Bihar 652. 2001; Kabir & Nair, Fl. Tamil Nadu Grasses 311. 2009. *Panicum pumilum* Poir. in Lam., Encycl. Suppl. 4. 273. 1816. *Setaria pallide-fusca* (Schumach.) Stapf. & Hobb. in Kew Bull. 1930: 259. 1930; Bor, Grass. Burm. Ceyl. Ind. Pak. 363. 1960. *Panicum pallide-fuscum* Schumach., Beskr. Guin. Pl. 58. 1827.

Annuals. Tufted, erect or sub-erect annual, 25-75 cm high. Leaf blades 4.8-20.1 x 0.3-0.8 cm, glabrous or sparingly puberulous, acuminate to acute at apex; sheaths glabrous, lower compressed and keeled, upper striate; ligule as a ciliate rim. Inflorescence 1.6-6.1 cm long, erect, dense cylindric false spike; rachis puberulous; branches reduced to a sessile involucral bristle, supporting as solitary perfect spikelet, frequently accompanied by an arrested one; pedicels reduced to small stumps with a discoid tip; bristles 6-12, up to 8 mm long, pinkish and brownish, slender, rigid. Spikelets 1.9-2.5 x 1.2 mm. Lower glume 3-nerved, ovate-lanceolate. Upper glume male slightly shorter than the spikelets, 5-nerved, broadly lanceolate, acute. Lower floret: male or rarely barren. Lemma up to 2 x 1 mm, 5-nerved, acute. Palea equally the lemma, hyaline, broadly elliptic. Upper floret: lemma slightly shorter than the floret, transversely rugose, not keeled. Palea granular-punctate. Caryopsis 1.8-2.1 x 0.6-0.9 mm, rugose planoconvex.

Ecology: Common on grassy places, hedges, thickets along the paddy field and river banks.
Fl. & Fr.: Almost all round the year.
Distribution: Pantropic.
Specimens examined: Muria, 1010; Silodhar, 1056.

4. Setaria verticillata (L.) P. Beauv., Ess. Agrost. 51: 178. 1812; Hook.f., Brit. India 7: 80. 1896; Prain, Bengal Pl. 2: 1170. 1903; Haines, Bot. Bihar & Orissa pt. 5: 989. 1924; Bor, Grass. Burma Ceylon India & Pakistan 365. 1960; Singh et al., Fl. Bihar 652. 2001; Paria & Chattopadhyay, Fl. Hazaribagh District 2: 1131. 2005; Kabir & Nair, Fl. Tamil Nadu Grasses 312, 313. 2009. *Panicum verticillatum* L., Sp. Pl. ed. 2, 82. 1762.

Annual; culms tufted, ascending or erect, up to 0.3 m tall. Leaf-blade sparsely pilose, 2.8-30.2 x 0.5-2.4 cm, acuminate, at apex, rounded to narrowed down at base; sheaths sparsely hairy, keeled upwards; ligule truncate ciliate. Spiciform panicle. 2.4-12.1 cm long, long-peduncled, erect, compact, coarsely bristle; branches spirally arranged, terminating in a retrorsely barbed bristled. Spikelets 2.1-2.4 x 1.1-1.3 mm, pale green, broadly ovate, acute. Lower floret: barren. Lemma 2-2.6 x 1.3-1.8 mm, 5-7-nerved, broadly ovate, acute. Palea 0 or small, hyaline, up to 1.5-0.4 mm. Upper floret: broadly oblong. Lemma 1.8-2.1 x 1-1.2 mm, sub-membranous. Caryopsis 1.7-1.9 mm long finely rugose.

Ecology: Common on moist grassy places and shady damp localities.
Fl. & Fr.: Almost throughout the year.
Distribution: Generally in temperate and tropical regions.
Specimens examined: Pathalgara, 6012; Murtiakalan, 6075.

44. SPOROBOLUS R. Br.

Sporobolus diander (Retz.) P. Beauv., Ess. Agrost. 26, 147 & 178. 1812; Hook.f., Fl. Brit. India 7: 247. 1896; Prain, Bengal Pl. 2: 1213. 1903; Haines, Bot. Bihar & Orissa pt. 5: 973. 1923; Bor, Common Grasses of the United Provinces 202. 1940; Bor, Grass. Burma Ceylon India & Pakistan 629. 1960; Paria & Chattopadhyay, Fl. Hazaribagh District 2: 1133. 2005. *Agrostis diander* Retz. Obs. Bot. 5: 19. 1789 ('diandra'). *Sporobolus indicus* (L.) R. Br. var. *diander* (Retz.) Jov. & Gued. in Taxon 22: 163. 1973.

Local name: Ciriya-ka-dana.

Slender erect tufted annual; culms 20-80 cm high, smooth glabrous. Leaves 7-20 x 0.2-0.3 cm, glabrous, flat, tapering at apex; sheaths striate; ligule as a rim of hairs. Panicle pyramidal. Spikelets up to 1.4 mm long. Glume nerveless; lower 0.4-0.5 mm long, ovate, truncate; upper up to 0.9 x 0.4 mm long, ovate-lanceolate. Lemma up to 1.4 x 0.8 mm, faintly 1-nerved, ovate. Palea as long as the lemma. Lodicules minute (Illus. Plate-XXV; Fig.-27 and Plate-XXXXV; Fig.-66A and B).

Ecology: Common along the river-banks and in moist pastures.
Fl. & Fr.: All round the year generally; but mainly in the rainy season.
Distribution: Pantropic.
Specimens examined: Duragara, 1012; Kabilas, 1025; Dhoria, 1077.

45. THEMEDA Forsk.

Themeda quadrivalvis (L.) O. Kuntze, Rev. Gen. Pl 2: 794. 1891; Haines, Bot. Bihar & Orissa pt. 5: 1050. 1924; Bor, Grass. Burma Ceylon India & Pakistan 252. 1960; Singh et al., Fl. Bihar 657. 2001; Paria & Chattopadhyay, Fl. Hazaribagh District 2: 1134. 2005; Kabir & Nair, Fl. Tamil Nadu Grasses 481. 2009. *Andropogon quadrivalvis* L. in Murr., Syst. Veg. ed. 13, 758. 1774; *Anthistiria iliate* L. f., Suppl. 113. 1781; Hook. f., Fl. Brit. India 7: 213. 1896; Prain, Bengal Pl. 2: 1207. 1903.

Local name: Gunkar.

Annuals. Culms 30-90 cm tall, erect or ascending. Leaves 15-30 cm long; sheaths loose, glabrous; ligules membranous. Inflorescence usually a dense panicle; spatheoles cymbiform, 1.7-2.5 cm long, acuminate. Involucral spikelets 4.5-5.5 mm long, sharply acute. Lower glume 4.2-5 mm long, with golden brown tubercled hairs; upper glume equal to the lower, 3-nerved, glabrous. Lemma hyaline, 1-nerved, epaleate. Sessile spikelets bisexual, ca 4.5 mm long; callus hairy. Lower glume ca 4 mm long, 6-7 nerved; upper glume similar. Upper lemma with ca 0.5 mm long awn. Anthers 3, yellow, ca 1.5 mm long. Caryopsis brown, subcylindrical, 5-6 mm long; callus hairy.

Ecology: Common in fields, sandy and grassy grounds, roadsides, etc.
Fl. & Fr.: Sept.-Oct.
Distribution: Throughout India, Nepal, Pakistan.
Specimens examined: Bukar, 1107; Pathalgarwa, 6062.

46. THYSANOLAENA Nees

Thysanolaena maxima (Roxb.) Kuntze, Revis. Gen. Pl. 2: 794. 1891; Bor, Grass. Burm. Ceyl. Ind. Pak. 650. 1960; SreeKumar & Nair, Fl. Kerala Grasses 444. 1991; Singh et al., Fl. Bihar 659. 2001; Paria & Chattopadhyay, Fl. Hazaribagh District 2: 1135. 2005. *Agrostis maxima* Roxb., Fl. Ind. (Carey & Wallich ed.) 1: 319. 1820. *Thysanolaena agrostis* Nees in Edinb. New Phil. J. 18: 180. 1835; Hook.f., Fl. Brit. India 7: 61. 1896; Haines, Bot. Bihar & Orissa pt. 5: 982. 1924.

Reed-like perennial grass. Culms terete, 1-2.5m tall, erect. Leaves lanceolate, upto 30x4 cm, coriaceous, cordate, semi-amplexicaul at base, acuminate at apex; sheaths compressed, mouth hairy; ligules short, membranous. Panicle effuse, decompounds with filiform branches, upto 90 cm long. Spikelets 1-2 nate, greenish or purplish, 2-flowered, elliptic-lanceolate, upto 1.5 mm long, acuminate, ciliate. Lower glume and upper long, oblong, acute, 1-nerved, empty, epaleate; upper lemma membranous, shorter than lower lemma, oblong, clothed with hairs; palea minute. Caryopsis minute.

Ecology: Common in shady slopes and damp banks along ravines and water courses.
Fl. & Fr.: Apr.-June.
Distribution: Throughout India.
Specimens examined: Ahri, 7002; Pathalgarwa, 7095.
Uses: As a fodder and brooms.

47. TRITICUM L.

Triticum aestivum L., Sp. Pl. 85. 1753; Bor, Grass. Burm. Ceyl. Ind. Pak. 679. 1960; Singh et al., Fl. Bihar 660. 2001; Paria & Chattopadhyay, Fl. Hazaribagh District 2: 1137. 2005; Kabir & Nair, Fl. Tamil Nadu Grasses 95. 2009. *T. vulgare* Vill., Hist. Pl. Dauph. 2: 153. 1787; Hook. f., Fl. Brit. India 7: 367. 1896. *T. sativum* Lam., Fl. Franc. 3: 625. 1778; Haines, Bot. Bihar & Orissa pt. 5: 693. 1924 (Repr. ed., 3: 1009. 1961).

Local name: Gehun, Gahum.

A tall annual grass. Leaves 13-35 x 0.5-1.2 cm, linear or linear-lanceolate, acuminate, flat, scaberulous on the veins. Spikelets erect, spicate, florets 3 or more. Bisexual floret: Gl. I cymbiform, 0.9-1 cm long, coriaceous, shortly awned. GI. II similar to GI. I. Lemma 0.9-1 cm long, strongly nerved, apex somewhat 3-toothed, the central one elongates as awn, awn often 1.4-7.5 cm long, spinulose; palea 8-8.5 mm long, 2-keeled, 2-nerved. Lodicules 2-2.5 mm long. Grain oblong. Neuter floret stalked.

Ecology: Cultivated as a winter crop and sometimes found as an escape in waste places, fields, etc.
Fl. & Fr.: Jan.-May.
Distribution: Cultivated in almost all parts of the world.
Specimens examined: Asnachuan, 6013; Khairtanr, 6025.
Uses: Grains are used in the form of ata, maida, suji, etc. It is an important food of India. The straw is also used for live-stock.

48. VETIVERIA Bory

Vetiveria zizanioides (L.) Nash in Small, Fl. Southeast U.S. 67. 1903; Haines, Bot. Bihar & Orissa pt. 5: 1032. 1924; Blatter & McCann, Bombay Grasses 65. 1935; Bor, Common Grasses of the United Provinces 216. 1940; Bor, Grasses Burm. Ceyl. Ind. Pak. 258. 1960; Cope in Nasir & Ali, Fl. Pakistan 143, 306. 1982; SreeKumar & Nair, Fl. Kerala Grasses 203. 1991; Singh et al., Fl. Bihar 661. 2001; Kabir & Nair, Fl. Tamil Nadu Grasses 488. 2009. *Phalaris zizanioides* L., Mant. 183. 1771. *Andropogon squarrosus* Hook. f., Fl. Brit. India 7: 186. 1896 (non L. f. 1781); Prain, Bengal Pl. 2: 1204. 1903 (Rep. ed. 2: 907. 1963).

Local name: Khas.

Perennials. A densely tufted grass up to 180 cm high; rhizome aromatic. Culms stout and rigid, glabrous. Leaves 30-90 x 0.4-1.4 cm, linear, acute, pale green; sheath striate, glabrous; ligule a scarious rim. Panicle 15-30 cm long, oblong or pyramidal, composed or racemes. Sessile spikelets linear, longer than the pedicelled ones. Gl. I 4 mm long, coriaceous, acute. Gl. II 3.5 mm long, cymbiform, 1-nerved. Lower lemma 2.5-3 mm long, empty, lanceolate, 2-nerved. Upper lemma bisexual, minutely 2-toothed; palea minute. Pedicelled spikelets linear-lanceolate. Pedicel 3.5 mm long, finely scabrid. Upper lemma entire, acute, usually male (Illus. Plate-XXVI; Fig.-28).

Ecology: Commonly grows on wet low lying ground, along banks of streams and rivulets. A good fodder when young. The dried roots are made into mats and hung over windows during summer.
Fl. & Fr.: Aug.-Dec.
Distribution: Almost throughout India.
Specimens examined: Chordaha, 6011; Sikda, 6024.
Uses: Roots are aromatic and are the source of vetiver oil, an ingredient in perfumes; also woven into fragrant mats, rugs and fans. Employed medicinally and also to control soil erosion.

49. ZEA L.

Zea mays L., Sp. Pl. 971. 1753; Hook.f., Fl. Brit. India 7: 102. 1896; Prain, Bengal. Pl. 2: 1209. 1903; Haines, For. Fl. Chota Nagpur 564. 1910 & Bot. Bihar & Orissa pt. 6: 1065. 1924; Bor, Grass. Burma Ceylon India & Pakistan 270. 1960; Singh et al., Fl. Bihar 661. 2001; Paria & Chattopadhyay, Fl. Hazaribagh District 2: 1139. 2005; Kabir & Nair, Fl. Tamil Nadu Grasses 489. 2009.

Annuals, monoecious. Culms 1-2 m tall, erect, glabrous. Leaves 25-90 x 5-10 cm, glabrous, sheaths terete; ligules 2-6 mm long, membranous. Male inflorescence a terminal panicle; rachis pubescent. Spikelets in pairs, up to 12 mm long. Anthers 3, yellow, ca 5 mm long. Female inflorescence a spadix, enclosed in large bracts. Spikelets in several rows along the axis (variable with varieties). Lower floret brown. Palea present or absent. Caryopsis of variable colour (pale yellow, purple, white), subglobose or dorsally flattened, ca 8 mm in diameter (Plate-XXXXV; Fig.-67).

Ecology: Cultivated throughout the state.
Fl. & Fr.: May-Oct.
Distribution: Almost throughout India.

Specimens examined: Danua, 6046; Chordaha, 6077.
Uses: Grains are used in the form of popcorn, sattu, etc. Immature cobs are eaten after roasting. Plant is also used in the form of a good fodder for cattle.

Analysis

There is no unanimity of opinion regarding the circumscription of a number of grass genera and even species. This has lead to different figures regarding the total number of world species and genera of grasses.

According to Clayton and Renvoize (1989) the total number of grass species in the world is about 10000 and these come under about 651 genera. Based on the number of genera it is the third largest family after Asteraceae and Orchidaceae. Species-wise its place is fifth after Asteraceae, Orchidaceae, Leguminosae and Rubiaceae (Clayton and Renvoize, 1989).

In India Poaceae is represented by 1334 species under about 261 genera (Karthikeyan, 1989). This works out to be about 14% of the total grass species of the world. Out of them 1184 are grasses under 241 genera, the rest being bamboos. In addition, there are 148 subspecies and varieties making the total member of infrageneric taxa of grasses available in India to 1332 (Kabeer and Nair, 2009). Thus in India its species-wise position is first closely followed by Orchidaceae (1229 species), Leguminosae (1192 species), Asteraceae (880 species), Rubiaceae (616 species) and Poaceae (545 species).

Haines (1924) publication "The Botany of Bihar and Orissa" was the first monumental work and it lists and describes almost all the members of Cyperaceae and Poaceae found in the state. However due to biotic and abiotic interferences, the floristic composition of certain regions has considerably changed. Singh et al. (2001) in the analysis of Flora of Bihar includes enumeration of Cyperaceae and Poaceae. They pointed out that Poaceae with 342 species is the largest family in the state, whereas *Cyperus* with 35 species is the largest genus. Similarly Paria and Chattopadhyay (2005) in Flora of Hazaribagh District have pointed out that Poaceae with 57 genera and 93 species starts first in term of the dominant family of Hazaribag District. On the other hand Cyperaceae with 41 species stands 4th in the rating of dominance whereas *Cyperus* with

15 species is the biggest genus of this District. The frequency and synopsis of the sedges of Gautam Buddha Wildlife Sanctuary, Hazaribag have been given (Table-5, 6 and Graph-1).

In the present study of the sedges of Gautam Buddha Wildlife Sanctuary, it is discerned that the family Cyperaceae is represented by 9 genera as opposed to 12 that are reported in Flora of Hazaribagh District (Paria and Chattopadhyay, 2005). 20, Flora of Bihar (Singh et al., 2001) and 11, Botany of Bihar and Orissa (H.H. Haines, 1924) Table-7 and Graph-2. After collating the data it was observed that the Genus *Bulbostylis* has two species namely *B. barbata* and *B. densa*. Flora of Hazaribagh District reported only one species. Flora of Bihar and Botany of Bihar and Orissa reported 3 species each. The genus *Carex* has only one species in Gautam Buddha Wildlife Sanctuary namely *Carex cruciata*. Whereas Flora of Hazaribagh District, Flora of Bihar, Botany of Bihar and Orissa 1, 10 and 6 species respectively. *Cyperus* is the largest genus in the sense that it has 15 species in Gautam Buddha Wildlife Sanctuary and 15 (P), 35 (S), 32 (H) in the other three respectively. The second largest representative genus is *Fimbristylis* with 6 (G), 10 (P), 24 (S) and 23 (H) species (Table-7). The genus *Mariscus, Pycreus* and *Schoenoplectus* have same number of species i.e. 2 in (G), in (P) they are represented by 2, 2 and 3 species respectively. In (S) they have 7, 7 and 11 and in (H) they have 6 and 7 in *Mariscus* and *Pycreus* while *Schoenoplectus* at the time included in *Scirpus* with 8 spp. *Kyllinga* has 3 each in (G) and (P) while in (S) and (H) it has 6 and 4 respectively. The genera which have been reported in (P) but not in (G) are *Eleocharis, Rikliella* and *Scleria*. The facts gleaned from the data are that altogether 34 spp. and 2 ssp. have been reported from Gautam Buddha Wildlife Sanctuary. This is not much less than that reported by Paria and Chattopadhyay (2005) in Flora of Hazaribagh District. Singh et al. (2001) reported altogether 139 species, 3 sub species and 20 varieties. Haines (1924) reported 100 species and 4 varieties.

Altogether 12 sedges reported from here (Gautam Buddha Wildlife Sanctuary) and not reported from Hazaribag by Paria and Chattopadhyay (2005) are *Bulbostylis densa, Cyperus bulbosus, C. castaneus, C. dubius, C. nutans, C. pangorei, C. pygmaeus, Fimbristylis alboviridis, F. littoralis, F. polytrichoides, Mariscus compactus* and *M. paniceus*. Out of them *Cyperus bulbosus, Fimbristylis alboviridis, F. littoralis* and *F. polytrichoides* were not reported by Haines (1924) Table-8. The species not reported by Singh et al. (2001) are *Cyperus dubius, Fimbristylis alboviridis* and *F. polytrichoides*.

Cyperaceae is an enigmatic family in the sense that the genera and species have too many overlapping characters and as such have been treated differently by various taxonomists; resulting into having synonyms altered for to often and citations lengthening with the passing time. The genus *Pycreus* has been treated here as separate from *Cyperus* based on the shape of the nuts which are bilaterally flattened with one angle facing the rachilla. The nut being trigonous in *Cyperus*. *Pycreus* separated from *Cyperus* on the basis of its 2 – fid style and flattened nut as approved to the 3 – fid style and triangular as dorsiventrally flattened nut of *Cyperus* proper. Hooker (1896), Haines (1924) and Fischer in Gamble (1931) also maintained a separate identity for *Pycreus*. Many workers however have amalgamated *Pycreus* with *Cyperus* based upon a broad and common floral character of their glumes being distichous. The two species of *Pycreus* of Gautam Buddha Wildlife Sanctuary are *P. flavidus* and *P. pumilus*.

The genus *Schoenoplectus* has been incorporated in *Scirpus* by quite a number of scientists. On the other hand many have maintained a separate identity for the genus *Schoenoplectus* (Paria and Chattopadhyay, 2005). The genus being represented by the species *S. articulatus* and *S. supinus* in Gautam Buddha Wildlife Sanctuary. The two monotypic genera found in Gautam Buddha Wildlife Sanctuary are *Carex* (*C. cruciata*) and *Fuirena* (*F. ciliaris*). The genus *Bulbostylis* has two species namely *B. barbata* and *B. densa*. The later being cited as *B. capillaris* by H.H. Haines, 1924. The three spices of *Kyllinga* found in Gautam Buddha Wildlife Sanctuary are *K. nemoralis, K. bulbosa* and *K. brevifolia*. Some of the floristic characters of *Kyllinga* closely resembles that of *Cyperus* thereby leading to the renaming of *Kyllinga species* as *Cyperus*. One of the three species here *K. brevifolia* has been assigned the name of *Cyperus brevifolius* (Rao and Verma, 1972). However Paria and Chattopadhyay (2005) and Singh et al. (2001) followed Haines (1924) and held it as *K. brevifolia*.

Mariscus is also closely aligned to *Kyllinga*. The unifying character of the two being their (a) glumes distichously arranged and (b) articulate rachilla. However the character which separates them is the 2 – fid style of *Kyllinga* as apposed to the 3 – fid style of *Mariscus*. The genus *Mariscus* has two species in Gautam Buddha Wildlife Sanctuary, i.e. *M. compactus* and *M. paniceus*. Due to its overlapping characters *M. paniceus* was called as *K. panicea* by Rottb. (1773); however Singh et al. (2001) kept it as *M. paniceus*. Some of the perennial and uncommon sedges of Gautam Buddha Wildlife Sanctuary are *Carex cruciata*, *Cyperus bulbosus*, *Cyperus dubius*, *Cyperus pygmaeus* and

Fimbristylis polytrichoides. Majority of the Cyperaceae are instead common and uniformly distributed (Table-9).

The family Poaceae being the bigger among the two has under representation. Some of the uncommon grasses are *Apluda mutica* L. which has been reported from Mainukhar and Sanjha villages, *Chloris dolichostachya* from Murtiakalan and Bukar, *Coix lachryma-jobi* from Ahri and Chordaha, *Dichanthium ischaemum* from *Morainia* and *Asnachuan, Digitaria longifolia* from Sanjha and Murtiakalan. *Eragrostis ciliata* from Pathalgara and Duragara and *Hackelochloa granularis* from Kabilas and Morainia village (Table-10). Although the plants in study could be found almost throughout the area, there are areas where they enjoyed maximum spread (Table-11). Village Chordaha has the maximum concentration of the plants. Danua, Garmorwa, Bukar and Silodhar were the other village which had thick pockets of grasses and sedges. Villages Asnachuan, Dhoria, Kabilas, Khairtand, Murtiakalan, Ahri and Morainia were some villages which had moderate representation of the plants. Places which showed had little grasses and sedges were Chamargadda and Kenduadih.

No comprehensive account on the grass flora of this state is available. Hooker (1896) in his Flora of British India treats about 700 species in about 130 genera of grasses. Grasses of Burma Ceylon India and Pakistan (Bor, 1960) is the main standard reference book for the identification of Indian grasses. But as during the past four decades after its publication, considerable additional data on the taxonomy and other aspects of this group have accumulated. This work cannot be completely depended upon for naming grasses of this area. Grasses and Bamboos of India (Moulik, 1997) is the recent book on Indian Poaceae. However, this publication also does not give any description of taxa as in the case of the book mentioned above. Another important publication worth mentioning is Florae Indicae Enumeratio: Monocotyledoneae (Karthikeyan et al., 1989) which contains a detailed check list of Indian grasses. Detailed studies and the resultant well documented floristic accounts on the grasses of different parts of India like Kerala, North-East India, Maharashtra etc. have been published in recent years. Flora of Tamil Nadu – Grasses (Kabeer and Nair, 2009) is another recent book which has been referred during the present work.

Altogether 49 genera of Poaceae have been reported from Gautam Buddha Wildlife Sanctuary during this work. The total number of species reported is 76 (Table-10). The most dominant genus is *Eragrostis* with 7 spp. followed by *Panicum* and *Setaria* each having 4 spp. *Digitaria* and *Saccharum* have 3 spp. each. The other dominant genera are *Sacciolepis, Aristida, Bothriochloa, Brachiaria* and *Dicanthium* (Table-12). In confirmity to my findings Paria and Chattopadhyay, 2005 also marked *Eragrostis* (10 spp.) as the largest genus of Hazaribag District. *Panicum* (5 spp.), *Setaria* (4 spp.) and *Digitaria* (4 spp.) were also the other three major genera which again confirmed my findings. However *Cynodon* (3 spp.) is the 5th largest genus from Hazaribag has only species *C. dactylon* in Gautam Buddha Wildlife Sanctuary. The other genera that have good number of species in Hazaribag and which are also represented from Gautam Buddha Wildlife Sanctuary are *Dichanthium, Echinochloa, Paspalum, Oryza* and *Saccharum*. It was also found that *Eragrostis, Digitaria* and *Panicum* were the major genera of Bihar (Jharkhand) (Singh et al., 2001) and India (Moulik, 1997).

Altogether 13 (thirteen) rare grasses which have been reported from Gautam Buddha Wildlife Sanctuary and not by Paria and Chattopadhyay (2005) are *Aristida adscensionis, Bothriochloa pertusa, Brachiaria reptans, Chloris dolichostachya, Desmostachya bipinnata, Digitaria bicornis, Iseilema anthephoroides, Paspalidium geminatum, Sacciolepis interrupta, Sacciolepis myosuroides, Sehima nervosum, Setaria pumila* and *Vetiveria zizanioides*. Some of these grasses which were also reported by Haines (1924) from Hazaribag are *Aristida adscensionis, Sacciolepis myosuroides, Sehima nervosum* and *Vetiveria zizanioides* and the grasses which were not reported by him from Hazaribag (Table-13) are *Brachiaria reptans, Chloris dolichostachya, Desmostachya bipinnata, Iseilema anthephoroides, Paspalidium geminatum* and *Sacciolepis interrupta*.

The 3 grasses which have been reported from Gautam Buddha Wildlife Sanctuary but not by Haines (1924) are *Bothriochloa purtusa, Digitaria bicornis* and *Setaria pumila* (Table-13). Only some of these grasses where reported by Singh et al. (2001) from Hazaribag and the rest being located outside Hazaribag.

In course of study grasses were located which have great Ecological significance (Table-14). *Chloris barbata, Cynodon dactylon, Desmosta-chya bipinnata, Eleusine indica, Aristida adscensionis* and *Echinochloa colona* were found to be drought resistant. Some of the grasses were good soil binders of rivers and stream beds; they being *C. dactylon, D. bipinnata, E. indica, Panicum paludosum, P. psilopodium* etc. The grasses which fought soil erosion on hilly tracks are *C. dactylon, Cymbopogon martinii, D. aegyptium* etc. Most significant finding was identifying grasses which colonized left over mines dunes. Some of them being *Aristida setacea, Saccharum spontaneum, S. bengalense, Alloteropsis cimicina, Digitaria ciliaris* and *D. longifolia*.

Besides these, the present work also shows unreported sedges, the list of the cultivable grasses (Table-15) and also those which are found in dry condition (Table-16). Some grasses reappear after forest fire (Table-17) whereas some others are used for forestation (Table-18). Few members of Poaceae have been reported from the Walls (Table-19) while some are found in the crevices of rocks (Table-20). The present study shows that there are few grasses abound in waste lands (Table-21). In Hindu rituals, grasses are also used (Table-22). *Aristida* has 2 spp. reported from this Sanctuary while 7 spp. from Bihar by Singh et al., 2001. Frequency and Synopsis of the grasses have also been given (Table-23, 24 and Graph-3). In this present work list of the perennial sedges and grasses (Table-25) and which are found in or near water courses (Table-26) along with very common sedges and grasses have been given (Table-27). Different kinds of products are made from the members of Cyperaceae and Poaceae (Table-28). These are also used as fodders and weeds of farms and gardens (Table-29 and 30). There is also a list of sedges and grasses which are uncommon to the Sanctuary (Table-31). The present work also shows artificial key to the grasses found in Gautam Buddha Wildlife Sanctuary, Hazaribag (Table-32). It was found that the groups D – H were the major once and also 80% of the species belonged to these groups. Among these group E was the biggest one which has the principal character of a single inflorescence of raceme/spike/cylinder; the ideal representatives being *Heteropogon contortus, Imperata cylindrica* etc. The other significant group with typical delining character of inflorescence being terminal whorled or digitate is Group F; and the species included here are best exemplified by *Alloteropsis cimicina* and *Bothriochloa* spp. and *Dichanthium* spp. The monotypic groups are B and I having *Coix* sp. and *Cymbopogon* sp. respectively. In the present work 20% of the sedges and grasses have been found of amphibian character (Graph-4).

Summary

Gautam Buddha Wildlife Sanctuary constitutes a part of Wildlife Division, Hazaribag. It is located within Hazaribag District and has a forest area of 100.05 sq. km. It provides a variety of flora and fauna. It's nearest Railway station being Koderma (41 km) and nearest airport Ranchi (160 km).

Due to deforestation and mining activities in and around the area of my study i.e. Gautam Buddha Wildlife Sanctuary, Hazaribag, sedges and grasses account for the largest group of plant in this region. Before we perceive the loss due to the falling of trees, grasses come into the picture and we somehow or the other realize the perception of the flaw in the terrestrial scheme, by which what was good for God's birds was bad for God's garden or that cruelty towards one set of creation (here trees) was benevolence (here grasses) towards another. Within a short span the once thick forest cover are taken over by grasses and sedges and turned into a grazing ground (grassland). However, they have not been adequately studied so far by taxonomists. Therefore, the literature available is scanty.

Cyperaceae or the sedge family with its wide range of distribution and habitat adaptability is adequately represented in this area. On the other hand sedges are mostly perennial or sometimes annual herbs. Because many species have a tufted growth habit, long thin-texture, narrow flat leaves with a sheathing base, a jointed stem and much branched inflorescence of tiny flowers, they are often described as graminoid, meaning grass like. Indeed many horticultural references include sedges under the general heading of grasses. Further more, many sedges and grasses do not fit the graminoid image at all, for example having leaf blades rounded in cross section or no leaf blades at all, or having compact, head like inflorescences. So, we end up with some sedges and grasses being called graminoid, and others being simply called herbs. To minimize confusion we prefer to restrict the term graminoid to true grasses (Family Poaceae) and to apply the new term cyperoid to members of the sedge family.

Grasses are rather interesting in that they are usually successful in occupying large tracts of land to the exclusion of other plants. If we take into consideration the number of individuals of any species of grasses, they will be found to out-number those of any species of any other family. Even as regards the number of species this family ranks fifth, the first four places being occupied respectively by Compositae, Leguminosae, Orchidaceae and Rubiaceae. As grasses form an exceedingly natural family it is very difficult for beginners to readily distinguish them from one another.

In the present study of the sedges of Gautam Buddha Wildlife Sanctuary, it is discerned that the family Cyperaceae is represented by 9 genera as opposed to 12 that are reported in Flora of Hazaribagh District (Paria and Chattopadhyay, 2005); 20 in Flora of Bihar (Singh et al., 2001) and 11 in Botany of Bihar and Orissa (H.H. Haines, 1924). After collecting the data it was observed that the genus *Bulbostylis* has two species namely *B. barbata* and *B. densa.* Paria and Chattopadhyay (2005) reported 1 species. The genus *Carex* has only one species in Gautam Buddha Wildlife Sanctuary namely *Carex cruciata* where as Flora of Hazaribagh District, Flora of Bihar, and Botany of Bihar and Orissa reported 1, 10 and 6 species respectively. *Cyperus* is the largest genus in the sense that it has 15 species in Gautam Buddha Wildlife Sanctuary. The second largest representative genus is *Fimbristylis* with 6 reported from Sanctuary, 10 from Flora of Hazaribagh District, 24 from Flora of Bihar and 23 species from Botany of Bihar and Orissa. The genera *Mariscus, Pycreus* and *Schoenoplectus* have same number of species i.e. 2 in the Sanctuary; in Flora of Hazaribagh District, they are represented by 2, 2 and 3 species respectively. In Flora of Bihar, they have 7, 7 and 11 and in Botany of Bihar and Orissa, they have 6 and 7 in *Mariscus* and *Pycreus* while *Schoenoplectus* was at that time included in *Scirpus* with 8 spp. *Kyllinga* has 3 each in Gautam Buddha Wildlife Sanctuary, Hazaribag and Flora of Hazaribagh District while in Flora of Bihar and Botany of Bihar and Orissa, it has 6 and 4 respectively. It is imperative to note that the genera which have been reported in Flora of Hazaribagh District but not in Gautam Buddha Wildlife Sanctuary, Hazaribag are *Eleocharis, Rikliella* and *Scleria*. The facts gleaned from the data are that altogether 34 species and 2 sub-species have been reported from Gautam Buddha Wildlife Sanctuary. This is not much less than that reported by Paria and Chattopadhyay (2005) in Flora of Hazaribagh

District. Singh et al. (2001) reported altogether 139 species, 3 subspecies and 20 varieties. Haines (1924) reported 100 species and 4 varieties. Altogether 12 sedges which have been reported from Gautam Buddha Wildlife Sanctuary but not from Hazaribag by Paria and Chattopadhyay (2005) are *Bulbostylis densa, Cyperus bulbosus, C. castaneus, C. dubius, C. nutans, C. pangorei, C. pygmaeus, Fimbristylis, alboviridis, F. littoralis, F. polytrichoides, Mariscus compactus* and *M. paniceus*. Out of them *C. bulbosus, F. alboviridis, F. littoralis* and *F. polytrichoides* were not reported by Haines (1924). The species not reported by Singh et al. (2001) are *C. dubius, F. alboviridis* and *F. polytrichoides*.

Altogether 49 genera of Poaceae have been reported from Gautam Buddha Wildlife Sanctuary during this work. The total number of species reported is 76. The most dominant genera are *Eragrostis* with 7 species followed by *Panicum* and *Setaria* each having 4 spp. *Digitaria* and *Saccharum* have 3 spp. each. The other dominant genera are *Sacciolepis, Aristida, Bothriochloa, Brachiaria* and *Dichanthium*. In confirmity to my findings Paria and Chattopadhyay (2005) also marked *Eragrostis* (10 spp.) as the largest genus of Hazaribag District. *Panicum* (5 spp.), *Setaria* (4 spp.) and *Digitaria* (4 spp.) were also the other three major genera which again confirmed my findings. However *Cynodon* (3 spp.) which is the 5th largest genus from Hazaribag has only one species *C. dactylon* in Gautam Buddha Wildlife Sanctuary. Altogether 13 (thirteen) rare grasses which have been reported from Gautam Buddha Wildlife Sanctuary and not by Paria and Chattopadhyay (2005) are *Aristida adscensionis, Bothriochloa pertusa, Brachiaria reptans, Chloris dolichostachya, Desmostachya bipinnata, Digitaria bicornis, Iseilema anthephoroides, Paspalidium geminatum, Sacciolepis interrupta, S. myosuroides, Sehima nervosum Setaria pumila* and *Vetiveria zizanioides*.

In course of study grasses of great ecological significance have been reported. *Cynodon dactylon, Desmostachya bipinnata, Eleusine indica, Aristida adscensionis* and *Echinochloa colona* were found to be drought resistant. Some of the grasses were good soil binders. They are *C. dactylon, D. bipinnata, E. indica, Panicum paludosum, P. psilopodium* etc. The grasses which fought soil erosion on hilly tracks are *C. dactylon, Cymbopogon martinii, Dactyloctenium aegyptium* etc. Most significant finding was identifying grasses which colonized left over mines dunes. Some of them being *A. setacea, Saccharum spantaneum, S. bengalense, A. cimicina, Digitaria ciliaris* and *D. longiflora*.

The identification of the grasses on the basis of their vegetative parts is both unscientific and erroneous. It also goes against the inviolable rule that the basis of classification of the grasses is spikelet. However with certain degrees of dexterity and intelligence one may be trained to identify certain grasses which we frequently come along in my area of study. It is in this light that we prepared an Artificial Key (Bor, 1941) and assigned grasses found in Gautam Buddha Wildlife Sanctuary to 10 groups (A – J) Table-32. The Key has been kept as simple as possible and where it seems possible to make a mistake in relegating a species to one group or another, that species is included in both groups. All the same the key does not obviate the study of microscopic characters altogether and that the use of hand lens and microscopic study together with the study of macroscopic characters have been included in the preparation of the key. It was found that the groups D – H were the major once and also 80% of the species belonged to these groups. Among these group E was the biggest one which has the principal character of a single inflorescence of raceme/spike/cylinder; the ideal representatives being *Heteropogon contortus, Imperata cylindrica* etc. The other significant group with typical delining character of inflorescence being terminal whorled or digitate is Group F; and the species included here are best exemplified by *Alloteropsis cimicina* and *Bothriochloa* spp. and *Dichanthium* spp. The monotypic groups are B and I having *Coix* sp. and *Cymbopogon* sp. respectively.

The involvement of the governmental agencies in tackling ecological perils has long been overemphasized but total dependency on them is again foolhardy. Also the degree of dependence on NGOs or Local bodies in these efforts is a vexed problem. This has led to a situation where neither is ready to share responsibilities. This in turn has left our precious green cover to the mercy of contractors. Tired of official obfuscation on the government side, the self help groups if they ever exist have in recent years raised up their ante against the colosal loss of the green Pagoda. The self-help operation has a positive role to play. Criticizing the increasing apathy of the establishment is not going to alleviate the situation; rather it would be detrimental in obtaining their much needed favour and help. The grass lands are definitely not the panacea for the lost canopies but certainly they have done their bit to save the nature from cataclypse and catastrophe.

We have at least four reasons for wanting to keep grasses out of harm's ways:

- To avoid the denuded forest turning into barren lands. Somehow or the other grasses tend to survive and flourish in denuded tracts.

- Grasses sustain the animal wealth of the area. Both domestic and wild grazers are dependent upon them.
- Many grasses are excellent soil binders and they check the precious soil from being swept off due to rains, flood and air storms.
- Some of the grasses are proven hearth burners and bread earners. They are being used fairly commonly in house hold activities, and in cottage industries.

Thus it is a racing certainty that unless and until something is quickly done we will continue to witness the perils of unplanned growth. Taxonomists here have a role to play and study the neglected plants such as sedges and grasses are all the more important to replenish our knowledge base about them. Our floral wealth is being cornered with pressure mounting on it from within the society and from without. It would indeed be an egregious mistake to take things lying down.

References

Achariar, K.R. and C.T. Mudaliar. 1921. *A handbook of South Indian Grasses.* Madras.

Alava, R.O. 1952. Spikelet variation in *Zea mays* L. *Ann. Missouri Bot. Gard.* 39: 65-96.

Ambasht, R.S. 1964. Ecology of the underground parts of *Cyperus rotundus* L. *Trop. Ecol.* 5: 67-74.

Ambasta, N. 2012. Taxonomy of *Cynodon dactylon* (L.) Pers., A Medicinal grass of Hazaribag, Jharkhand, India. *Sci. Res. Rept.* 2(1): 34-35.

Ambasta, N. and C.T.N. Singh 2010. Taxonomical Study on *Aristida setacea* Retz., a veritable treasure for tribals of Hazaribagh District. *Biospectra* 5(1): 83-84.

Ambasta, N. and N.K. Rana 2013. Taxonomical Study of *Chrysopogon aciculatus* (Retz.) Trin., a significant grass of Chauparan, Hazaribag, Jharkhand. *Sci. Res. Rept.*3(1): 27-29.

Ambasta, S.P. 1992. *The Useful plants of India*. New Delhi.

Anderson, D.E. 1961. Taxonomy and distribution of the genus *Phalaris. Iowa St. J. Sci.* 36: 1-96.

Anderson, D.E. 1974. Taxonomy of the genus *Chloris* (Gramineae). *BrighamYoung Univ. Sci. Bull. Biol. Ser.* 19 (2): 1-133.

Annamalai, P., P.J. Chandramohan and V. Srinivasan. 1962. An abnormal stem of *Oryza sativa L. Madras Agric. J.* 49: 314.

Archer, C. 2005. *Family Cyperaceae.* National Herbarium, Pretoria.

Ashalatha, V.N. and V.J. Nair. 1993. *Brachiaria Griesb.* and *Urochloa* P. Beauv. (Poaceae) in India a conspectus. *Bull. Bot. Surv. India* 35 (1-4): 27-31.

Avdulov, N.P. 1931. *Cytotaxonomic investigation in the family Gramineae.*

Baaijens, G.J. and J.F. Veldkamp. 1991. *Sporobolus* (Gramineae) in Malesia. *Blumea* 35 (2): 393-458.

Backer, C.A. and Backhuizien van den Brink Jr. 1968. *Flora of Java vol. 3. Gramineae.* Noordhoff, Netherlands.

Backer, C.A. and R.C.B. Brink. 1968. Cyperaceae' in *Flora of Java* 3: 451-495. Leyden.

Barnard, C. 1964. *Grasses and Grasslands.* Macmillan.

Bennet, S.S.R. 1987. *Name Changes in Flowering Plants of India and Adjacent Regions.* Dehra Dun.

Bentham, G. 1881. Notes on Cyperaceae with special reference to Lestiboudois 'Essai' on Beauvois Genera. *J. Linn. Soc. (Bot.)* 18: 360-368.

Bentham, G. and J.D. Hooker. 1862-1883. *Genera Plantarum*, Vol. 3. London.

Bews, J.W. 1929. *The world's Grasses.* Longmans, Green and Co. London.

Bhangale, J. and S. Acharya 2014. Antiarthritic activity of *Cynodon dactylon*(L.) Pers. *Indian Journal of Experimental Biology* 52: 215-222.

Bharadwaja, R.C. 1956. On the distribution and origin of the genus *Ischaemum. Sci. Cult.* 21: 748-749.

Bhattacharya, P.K. and Krishnendu Sarkar. 1998. *Flora of West Champaran District*, Bihar, B.S.I., Calcutta.

Bhattacharya, U.C. 1997. *Flora of West Bengal* Vol. I. B.S.I., Calcutta.

Bhimaya, C.P., N.D. Rege and S. Srinivasan. 1956. Prelimininary studies on the role of grasses in soil conservation in the Nilgiris. *J. Soil Water Cons. India* 4: 113-117.

Blatter, E. 1911. A bibliography of the botany of British India and Ceylon. *J. Bombay. Nat. Hist. Soc.* 20: 79-185.

Blatter, E. and C. McCann, 1934. Revision of the Flora of Bombay Presidency – 'Cyperaceae' in *J. Bombay. Nat. Hist. Soc.* 37: 15-35; Ibid. 37: 255-277. Ibid. 37: 532-548. 1934; Ibid. 37: 764-779. 1935; Ibid. 38: 6-18. 1935.

Blatter, E.J. and C. McCann. 1934. *The Bombay Grasses.* Sci. Monogr. No. 5. Imp. Counc. Agric. Res. Mem. India.

Boeckeler, O. 1868; Die Cyperaceen des Koeniglichen Herbariums *Zu Berlin. Linnaea* 35: 397-612; Ibid. 36: 271-512. 1870; Ibid. 36: 691-768. 1870; Ibid. 37: 129-142. 1872; Ibid. 37: 520-544. 1872; Ibid. 37: 545-647. 1873; Ibid. 38: 223-544. 1874; Ibid. 39: 1-152. 1875; Ibid. 40: 327-452. 1876; Ibid. 41: 145-356. 1877.

Boott, F. 1858. *Illustrations of the genus Carex* pt. I. tt. 1-200; pt. II. tt. 201-310. 1860; pt III. tt. 311-411. 1862; pt. IV. tt. 412-600. 1867. London.

Bor, N.L. 1938. Some remarks on the geology and flora of the Naga and Khasi hills. 150th *Anniv. Vol. Roy. Bot. Gard. Calcutta* 129-135.

Bor, N.L. 1940. *Flora of Assam.* Vol. 5. Gramineae. Calcutta.

Bor, N.L. 1941. *Some common U.P. Grasses.* Ind. For. Rec. n.s. (Bot.) 2 (1): 1-220.

Bor, N.L. 1948. Dr. Stocks' *Sporobolus* from Sind. *Kew Bull.* 1948: 45.

Bor, N.L. 1953-54. The genus *Cymbopogon* in India. *J. Bombay Nat. Hist. Soc.* 51 (4): 890-916; II. 52 (1): 149-183.

Bor, N.L. 1960. *The Grasses of Burma, Ceylon, India and Pakistan.* London. *Reprinted,* 1973. Germany.

Bor, N.L. 1970. Gramineae. In: Rechinger, K.H. (ed.). *Flora Iranica.* Akad. Druck., Austria.

Bridson, G.D.R. and E.R. Smith. 1991. *Botanico-Periodicum Huntianum/ Supplementum.* Hunt Institute for Botanical Documentation, Pittsburgh. (*Supplement to* 1968 *edition*).

Brown, R. Novae 1810. Cyperaceae. *Prodrmus florae Novae Hollandiae et Indulae van - Diemen.* London.

Brummitt, R.K. and C.E. Powell. 1996. *Authors of Plant Names.* (repr. ed.). Royal Botanic Gardens, Kew.

Burkill, I.H. 1909. First notes on *Cymbopogon martinii* Stapf. *Proc. J. Asiat. Soc. Bengal.* 5 n.s. 3: 87-93.

Burman, N.L.B. 1768. *Flora Indica.* Amsterdam.

Caius, J.F. 1935. The medicinal and poisonous sedges of India. *J. Bombay. Nat. Hist. Soc.* 38: 163-170.

Chandrasekaran, S.N. and S. Daniel. 1947. A short note on a case of double digitate inflorescence in *Dactyloctenium aegyptium* willd. *Curr. Sci.* 16: 386.

Chatterjee, D. 1947. Botany of wild and cultivated rices. *Nature* 160: 234-137.

Chavan, A.R. and S.D. Sabins. 1960. Cyperaceae from Mt. Abu. *J. Ind. Bot Soc.* 39: 27-29.

Chinnamani, S. 1975. Grassland types and its management in India with special reference to Tambaram flora and *Sehima – Dichanthium* grasslands of Tamil Nadu. *The Naturalist* Vol. I.

Chopra, R.N., I.C. Chopra and B.S. Verma. 1969. *Supplement to Glossary of Indian Medicinal Plants.* C.S.I.R., New Delhi.

Chopra, R.N., S.L. Nayar and I.C. Chopra. 1956. *Glossary of the Indian Medicinal Plants.* C.S.I.R., New Delhi.

Clarke, C.B. 1884. On the Indian species of *Cyperus*; with remarks on some others that specially illustrate the subdivisions of the genus. *J. Linn. Soc. (Bot.)* 21: 1-202.

Clarke, C.B. 1893. 'Cyperaceae' in J.D. Hooker, *Flora of British India* 6: 585-672. et Ibid. 6: 673-748. London.

Clarke, C.B. 1894. On certain authentic Cyperaceae of Linnaeus. *J. Linn. Soc. (Bot.)* 30: 299-315.

Clarke, C.B. 1898. On the Sub-subareas of British India, illustrated by the detailed distribution of the Cyperaceae in that Empire. *J. Linn. Soc. (Bot.)* 34: 1-146.

Clarke, C.B. 1903. 'Cyperaceae' in Forbes & Hemsley, *an enumeration of all plants know from China proper*. Ibid. 36: 202. Ibid. 36: 296. 1903; Ibid. 36: 297-319. 1904.

Clarke, C.B. 1904. *List of Carices of Malaya*. Ibid. 37: 1-16.

Clarke, C.B. 1908. New genera and species of Cyperaceae. *Kew. Bull. Add. Ser.* 8: 1-196.

Clarke, C.B. 1909. *Illustrations of Cyperaceae* tt. 1-114. London.

Clayton, W.D. 1966. Studies in Gramineae. 13. *Kew Bull.* 20: 449.

Clayton, W.D. 1972. Gramineae In: Hepper's *Flora of West Tropical Africa*. Ed, 2. Vol. 3. Pt. 2. London.

Clayton, W.D. 1974. Studies in the Gramineae XXXVII. Notes on the genus *Digitaria* Haller. *Kew Bull.* 29 (3): 517-523.

Clayton, W.D. 1975. The *Paspalum scrobiculatum* complex in tropical Africa. *Kew Bull.* 30 (1): 101-105.

Clayton, W.D. 1979. Notes on *Setaria* (Gramineae). *Kew Bull.* 33 (3): 501-509.

Clayton, W.D. and J.R. Harlan. 1970. The genus *Cynodon* L.C. Rich. in tropical Africa. *Kew Bull.* 24: 185-189.

Clayton, W.D. and S.A. Renvoize. 1982. *Flora of Tropical East Africa, Gramineae.* Rotterdam.

Clayton, W.D. and S.A. Renvoize. 1989. *Genera Graminum: Grasses of the World*. HMSO London.

Cook, T. 1908. *Flora of Bombay Presidency*. Vol. 2. Bombay.

Cooke, T. 1909. 'Cyperaceae' in *The Flora of the Presidency of Bombay*. 3: 364-421.

Cope, T.A. 1982. Poaceae. In: Nasir, E. and S.I. Ali (eds.). *Flora of Pakistan*. 143: 1-678. Karachi.

Cronquist, A. 1968. *The evolution and classification of flowering plants.* Thomas Nelson and Sons Ltd. London.

D' Alameida, J.F.R. and C.S. Ramaswamy. 1948. A contribution to the study of the ecological anatomy of the Indian Cyperaceae. *Bot. Mem. Univ. Bombay.* 1: 1-180.

Dabadghao, P.M. and K.A. Shankaranarayan. 1973. *The Grass Cover of India.* ICAR, New Delhi.

Das, H.P. 1950. *Geography of Assam.* New Delhi.

Davis, G.L. 1966. *Systematic Embryology of Angiosperms.*

De Wet, J.M.J. and J.R. Harlan. 1970. Biosystematics of *Cynodon* L.C. Rich. (Gramineae). *Taxon.* 19: 565-569.

De Wet, J.M.J., K.E. Prasad Rao, D.E. Brink and H.M. Mengesha. 1984. Systematics and evolution of *Eleusine coracana* (Gramineae). *Amer. J. Bot.* 71: 550-557.

Deshpande, M.B. and G.L. Shah. 1968. A new species of *Fuirena* from Gujrat, India. Ibid. 10: 239.

Deshpande, U.R. 1990 (1988). The genus *Heteropogon* Pers. (Poaceae) in India. *Bull. Bot. Surv. India.* 30 (1-4): 120-125.

Deshpande, U.R. and N.P. Singh. 1986. *Grasses of Maharashtra.* Mittal Publications, New Delhi.

Duthie, J.F. 1883. *A list of the grasses of N.W. India-Indigenous and cultivated.* Roorkee.

Duthie, J.F. 1886. *Illustrations of the indegenous fooder grasses of the plains of N.W. India.* Roorkee.

Duthie, J.F. 1888. *The fooder grasses of North India.* Roorkee.

Dutta, R.M. and J.N. Mitra. 1955. Common plants in and around Dacca. *Bot Soc. Beng. Special Publ.* no. 4. Calcutta.

Featherly, H.I. 1973. *Taxonomic Terminology of the Higher Plants.* Ames.

Fischer, C.E.C. 1931. 'Cyperaceae' in Gamble, Flora of Madras Presidency pt. IX: 1620-1687.

Fischer, C.E.C. 1934. *Flora of Madras.* Gramineae pt. X. pp. 1789-1864.

Fischer, C.E.C. 1934-1936. in Gamble, *Flora of the Presidency of Madras.* Part X and XI (B.S.I. reprint 1957, Calcutta).

Fosberg, F.R. 1976. Status of the name *Chloris barbata* (L.) Swartz In: Nomenclature. *Taxon* 25 (1): 176-178.

Gamble, J.S. 1896. The Bambuseae of British India. *Ann. Roy. Bot. Gard. Calcutta.* 7: 1-133. tt. 1-119.

Gandhi, S.S. 1968. Variation in spike number of star grass [*Dactyloctenium aegyptium* (Linn.) P. Beauv.] *Labdev* 6B: 255-257.

Good, R. 1953. *The Geography of the Flowering Plants.* ed. 2. Longmans, London.

Gordon – Gray, K.D. 1965. Studies in Cyperaceae in Southern Africa–I *J. South Afr. Bot.* 31: 137-143. pt. II. Ibid. 31: 285-291. 1965; pt. III. Ibid. 32: 129-140. 1965: pt. IV. Ibid. 32: 141-152. 1966.

Govindarajalu, E. 1966. Studies in Cyperaceae–II. *Bull. Bot Surv. India* 8: 352. pt. III. *Proc. Indian Acad. Sci. Hist. Soc.* 69: 246-249. 1972; pt. V. Ibid. 69: 159-164. 1972; pt. VIII. Proc. Indian Acad. Sci. 76: 181-193. 1972; pt. IX. Ibid. 78: 45-58. 1973; pt. XI. Ibid. 79: 160-172. 1974; pt. XII. Ibid. 80: 41-50. 1974; pt. XIII. Ibid. 82: 205-210. 1975; pt XVI. Ibid. 88: 229-242. 1979; pt. XVII. Ibid. 91: 43-53. 1982; pt XVIII. Ibid. 81: 187-196. 1982.

Guaglianone, E.F. 1980. Contribution al estudio del genero *Rhynchospora* Vahl (Cyperaceae) 2. *Darwiniana* 22: 499-509.

Gupta, B.K. and R. Jaffer. 1982. On the rarity and identity of some Indian species of the genus *Cymbopogon. Indian J. Forest.* 5 (1): 73-74.

Hackel, E. 1889. Monographica Andropogonearum. Paris (in Condole) *Monogr. Phan.* 6.

Haines, H.H. 1910. *A Forest Flora of Chotanagpur* including Gangpur and the Santal Parganas, Dehra Dun, India.

Haines, H.H. 1924. 'Cyperaceae' in *Botany of Bihar and Orissa* pt. V: 888-937. Allahabad.

Haines, H.H. 1924. *Botany of Bihar and Orissa*, Vol. 3. London.

Haines, R.W. 1971. Amphicarphy in East African Cyperaceae *Mitt. Bot. Staatssamml. Manchen* 10: 534-538.

Hajra, P.K., R.R. Rao, D.K. Singh and B.P. Uniyal (Ed.). 1995. *Flora of India*, Vol. 12 and 13. Botanical Survey of India, Calcutta.

Hall, J.B. 1973. The Cyperaceae within Nigeria-distribution and habitat. *Bot. J. Linn. Soc.* 66: 323-346.

Henrard, J. Th. 1927-1932. *A Critical Revision of the Genus Aristida.* Vol. 1-3. Leiden.

Henrard, J. Th. 1929. *A Monograph of the Genus Aristida.* Meded. Van's Rijksherb. Leiden.

Henrard, J. Th. 1950. *Monograph on the genus Digitaria.* Leiden.

Hitchcock, A. 1920. *Revision of North American Grasses. Contrib. U.S. Nat. Herb.* 22: 1-208.

Hitchcock, A. 1933. New Grasses from Kashmir *J. Wash. Acad. Sci.* 23: 134-136.

Hooker, J.D. 1854. *Himalayan Journals* Vol. 1-2. London.

Hooker, J.D. 1896-1897. *Flora of British India.* Vol. 7. London.

Hooper, S.S. 1972. New taxa, names and combinations in Cyperaceae for the Flora of west tropical Africa. *Kew Bull.* 26: 577-583.

Hubbard, C.E. 1954. *Grasses. A guide to their structure, identification, uses and distribution in the British Isles.* Penguin Books. Bungay, Suffolk.

Huchinson, J. 1959. *The Families of the Flowering Plants.* Vol. 2. Oxford Univ. Press. London.

Jain, S.K. 1961. A contribution to the bibliography of Grasses. *Proc. Nat. Acad. Sci. Ind.* 31 B. 361-382.

Jain, S.K. 1967. The genus *Cynodon* Rich. Ex Pers. in India. *Bull. Bot. Surv. India* 9: 134-151.

Jain, S.K., D.K. Banerji and D.C. Pal. 1975. Grasses of Bihar, Orissa and W. Bengal. *J. Bombay Nat. Hist. Soc.* 72: 758-773.

Jain, S.K., Doli Das and D.K. Banerji. 1972. Further Contribution to bibliography of grasses. *Bull. Bot. Surv. India* 14: 24-45.

Jain, S.P., Z. Abraham and H.S. Puri. 2000. Botanical description and taxonomy of Indian *Cymbopogon.* In: Kumar, S. et al. (eds). *Cymbopogon: The Aromatic Grass Monograph.* CIMAP, Lucknow.

Judd, I.B. 1979. *Handbook of Tropical Forage grasses.* New York, Garland.

Kabeer, K.A.A. and V.J. Nair. 2009. *Flora of Tamil Nadu - Grasses.* BSI.

Kanjilal, U.N., P.C. Kanji Lal, R.N. De and A. Das. 1934-1940. *Flora of Assam,* Vol. 1-5. Shillong.

Karthikeyan, S., S.K. Jain, M.P. Nayar and M. Sanjappa. 1989. *Florae Indicae Enumeratio Monocotyledonae. Fl. Ind. Ser.* 4. Calcutta.

Kaul, V.K. and S.K. Vats. 1998. Essential oil composition of *Bothriochloa pertusa* and phyletic relationship in aromatic grasses. *Biochem. Syst. Ecol.* 26: 374-386.

Kern, J.H. 1952. A neglected Indian species of *Cyperus. Reinwardtia* 1: 463-466.

Kern, J.H. 1961. Cyperaceae of Thailand (excl. *Carex*). *Reinwardtia* 6: 25-83, 145-164.

Kern, J.H. 1962. New look at some Cyperaceae mainly from the tropical standpoint. *Advanc. Sci.* 19. 78: 141-148.

Kern, J.H. 1974. 'Cyperaceae' in van Steenis, *Flora Malesiana,* ser I. 7: 435-753. Leyden.

Koyama, T. 1967. Iconographia Cyperacearum-I. *Phytologia* 15: 201-221. tt. 1-10; pt. II. Ibid. 17: 396-421. tt. 11-20. 1968; pt. III. Ibid. 20: 218-241.

Koyama, T. 1969. Zwei neue Cyperaceen aus China. *Willdenowia* 5/3: 489-493.

Koyama, T. 1970. Beitrage zur Cyperaceenflora von Ceylon. *Bot. Mag. Tokyo* 83: 184-92.

Koyama, T. 1971. Systematic interrelationships among Sclerieae, Lagenocarpeae and Mapanieae (Cyperaceae). *Mitt. Bot. Staatssamml. Miinchen* 10: 604-617.

Koyama, T. 1972. Cyperaceae-Rhychosporeae and Cladieae. *Mem. N.Y. Bot. Gard.* 23: 23-89.

Koyama, T. 1979. Studies in the Cyperaceae of Thailand-III. New and critical species of the Cariceae. *Bot. Mag. Tokyo* 92: 217-233.

Kuekenthal, G. 1909. 'Cyperaceae-Caricoideae' in Engler, *Das Pflanzenreich* IV: 20. heft 38: 1-824. Beriln.

Kumar S., K. Kumar, Navneet and S.S. Gautam. 2014. Antibacterial evaluation of Cyperus rotundus Linn. root extracts against respiratory tract pathogens. *African Journal of Pharmacology and Therapeutics.* 3(3): 95-98.

Kunth, C.S. 1837. *Enumeratio Plantarum* (Cyperaceae) – II. Stuttgart.

Kurz, S. 1875. Bamboo and its uses. *Ind. For* 1: 219-269, 335-336.

Lawrence, G.H.M., A.F. Gunther Buchheim, G.S. Daniels and H. Dolezal. 1968. *Botanico-Periodicum-Huntianum.* Pittsburg.

Lazarides, M. 1976. The genus *Eragrostiella* Bor (Poaceae, Eragrostideae). *Contr. Herb. Austr.* 22: 1-7.

Lazarides, M. 1979. *Aristida* L. (Poaceae, Aristideae) in Australia. CSIRO, Canberra.

Lazarides, M. 1980. The Genus *Leptochloa* Beauv. (Poaceae, Eragrostideae) in Australia and Papua New Guinea. *Brunonia* 3: 247-269.

Lazarides, M. 1980. *Tropical Grasses of Southeast Asia.* J. Cramer, Hirschberg.

Linnaeus, C. 1753. *Species Plantarum* ed. 1. et ed. 2. Stockholm.

Linnaeus, C. 1754. *Genera Plantarum* ed. 5. Stockholm.

Linnaeus, C. 1767 and 1771. *Mantissa Plantarum.* Stockholm.

Lisboa, J.C. (Mrs.) 1890-1893. Bombay grasses. *J. Bombay Nat. Hist. Soc.* 5: 116-131, 226-232, 337-349. 1890; 6: 189-219. 1891; 7: 357-390. 1892; 8: 1074-119. 1893.

Love, A., D. Love and M. Raymond. 1961. Cyto-taxonomy of *Carex* sect. *Capillares. Canad. J. Bot.* 35: 715-761.

Luck, P.E. 1979. *Setaria,* an important pasture grass. *Queensland Agric.* J. 105: 136-143.

Maheshwari, P. 1950. *An introduction to the Embryology of Angiosperms.* New York.

Majumdar, A.M., V.P. Rao and S.O. Deshmukh. 1986. Occurrence of aflatoxin in *Paspalum scrobiculatum* L. seeds. *Indian J. Appl. Biol.* 1: 33-35.

Majumdar, R.B. 1973. The genus *Panicum* Linn. in India. *Bull. Bot. Soc. Bengal* 27: 39-54.

Majumdar, R.B. 1980. Indian grasses-nature, composition and classifica-tion. *Indian Biol.* 12: 44-57.

Malick, K.C. and R. Prasad. 1968. Notes on the identity and distribution of *Cyperus alulatus* Kern in India. *Indian Forester* 94: 885.

Matthew, K.M. 1982. *Illustrations on the Flora of the Tamil Nadu Carnatic* 2: 800-960. Tiruchirappalli.

McNeill, J., F.R. Barrie, H.M. Burdet, V. Demoulin, D.L. Hawksworth, K. Marhold, D.H. Nicolson, J. Prado, P.C. Silvia, J.E. Skog, J.H. Wiersema and N.J. Turland. 2006. *International Code of Botanical Nomenclature. (Vienna Code).* A.R.G. Gantner, Verlag KG.

Mehra, K.L. 1962. The *Dichanthium annulatum* complex. 1: morphology. Phyton 18: 87-93.

Mehra, K.L. 1963. Consideration on the African origin of *Eleusine coracana* (Linn.) Gaertn. *Curr. Sci.* 32: 300-301.

Micheli, P.T. 1729. *Nova Plantarum genera juxta Tournafortii diposita*. Florence.

Miquel, F.A.W. 1855-1859. Flora van Nederlandsch Indie-alternative title, *Flora Indiae Batavae,* Vol. 1-3. Amsterdam.

Mishra, A.K., A.K. Singh and S.K. Sharma. 2010. *Management Plan of Gautam Buddha Wildlife Sanctuary, Hazaribag.* (2008-09 to 2017-18).

Mooney, H.F. 1941. Some additions to the Botany of Bihar and Orissa. *Indian For. Rec.* (n.s.) 3: 63-119.

Mooney, H.F. 1950. *Supplement to the Botany of Bihar and Orissa*. Ranchi.

Moulik, S. 1997. *The Grasses and Bamboos of India.* Vol. 1 and 2. Jodhpur.

Mukerjee, S.K. 1945. A Botanical tour in Chotanagpur. *Bull. Bot. Soc. Bengal.* 1: 27-28.

Mukherjee, S.K. 1949. Studies on *Saccharum spontaneum* and allied Grasses. I: preliminary report on collection. *Indian J. Genet.* 9: 47-58.

Mukherjee, S. K. 1954. Revision of the genus *Saccharum. Bull. Bot. Surv. Bengal* 8: 143-148.

Mukherjee, S. K. 1957. Origin and distribution of *Saccharum. Bot. Gaz.* 119: 46-61.

Murty, U.R. 1975. Taxonomy of the genus *Apluda* Linn. *Bull. Bot. Surv. India* 14 (1-4): 149-151.

Naithani, H.B. and M.B. Raizada. 1977. Notes on distributional records on grasses. *Indian Forester* 103 (8): 513-524.

Nayak, N.C. and D.B. Misra. 1962. Cattle poisoning by *Paspalum scrobiculatum* (Kodua poisoning). *Indian Vet. J.* 31: 501.

Nees Von Esenback. 1834. Cyperaceae Indiacae praecipue juxta herbaria wightii, wallichii, Roylei *et* Lindleyi in Wight, R., *Contributions to the Botany of India*. London.

Nooteboom, H.P. 1978. A taxonomic revision of the Malesian and Australian species of *Uncinia* (Cyperaceae). *Blumea* 24: 511-520.

Panigrahi, G. 1965. Studies in the Monocot Flora of Assam and N.E.F.A. *Proc. Nat. Acad. Sci.* sect. B. 35: 357-366.

Paria, N.D. and S.P. Chattopadhyay. 2005. *Flora of Hazaribagh District, Bihar.* Vol. 2. BSI.

Parker, R.N. and W.B. Turill. 1929. 'Cyperaceae' in Duthie, *Flora of the Upper Gangetic Plain and of the adjacent Siwalik and sub-Himalayan tracts* 3: 320-371.

Patra, J.K., R.R. Mishra, S.D. Rout and H.N. Thatoi. 2011. An Assessment of Nutrient Content of Different Grass Species of Similipal Tiger Reserve, Orissa. *World Journal of Agriculture Sciences* 7 (1): 37-41.

Patunkar, B.W. 1980. *Grasses of Marathwada.* Scientific Publishers, Jodhpur.

Pax, F. 1888. 'Cyperaceae' in E. and P. *Die Naturlichen Pflanzen-familien* II. 2: 98-126.

Pradhan, S.G. 1985. A note on two interesting forms of *Digitaria bicornis* (Lamk.) Roem. and Schult. ex Loud. – Poaceae. *J. Econ Taxon. Bot.* 6 (2): 347-349.

Prain, D. 1903. *Bengal Plants.* Vol. 2. London.

Prasanna, P.V. and T. Pullaiah. 1988. Genus *Eragrostis* Wolf (Poaceae) in Andhra Pradesh. *J. Swamy Bot. Club.* 5 (1): 45-49.

Pyrah, G.L. 1969. taxonomic and distributional studies in *Leersia* (gramineae). *Iowa St. J. Sci.* 44: 215-270.

Raizada, M.B. 1948. Name changes in common Indian plants. *Indian For.* 84: 467-538.

Raizada, M.B. 1950. Review of some name changes in Indian grasses. *Indian Forester* 76: 523-525.

Raizada, M.B. 1959. Name changes in common Indian grasses. *Indian For.* 85: 437-509.

Raizada, M.B. 1966. Nomenclatural changes in Indian plants. *Indian For.* 92: 299-339.

Raizada, M.B., R.C. Bharadwaja and S.K. Jain. 1957. Grasses of Upper Gangetic plain. Panicoideae. Part I. *Indian For. Rec.* 4: 171-277.

Rao, A.S. 1974. The vegetation and phytogeography of Assam-Burma in Mani, *Ecology and Biogeography in India*. Hague.

Rao, A.S. and D.M. Verma. 1972. Materials towards a Monocot Flora of Assam – I (Hydrocharitaceae and Burmanniaceae). *Bull. Bot. Surv. India* 12: 139-143; pt. II (Zingiberaceae and Marantaceae). Ibid. 14: 114-143. 1975; pt. III (Taccaceae, Dioscoreaceae and Stemonaceae). Ibid. 15: 189-203. 1976; pt. IV (Pontederiaceae, Xyridaceae and Commelinaceae). Ibid. 16: 1-20. 1977; pt. V (Flagellariaceae, Juncaceae, Typ-haceae, Sparganiaceae, Araceae, Lemnaceae, Triuridaceae, Alismatace, Butomaceae, Aponogetonaceae, Potamogeto-naceae and Eriocaulaceae). Ibid. 18: 1-40. 1979.

Raymond, M. 1951. Sedges as materials for phytogeographical studies. *Mem. Jard. Bot. Montreal* 20: 1-24.

Raymond, M. 1955. Cyperacees d'Indo-chine–I. *Le Natur. Canad.* 82: 145-165.

Raymond, M. 1965. Cyperaceae novae vel criticae–IV. Some Cyperaceae from the Karakoram Range (Kashmir). *Le Nat. Canad.* 92: 76-80.

Raymond, M. 1966. Cyperaceae. *Studies in the Flora of Thailand – 39. Dansk Bot. Ark.* 23: 313-374.

Retzius, A.J. 1786/87. *Fasciculus Observationum Botanicarum* fasc. 4; fasc. 5. 1788; fasc. 6. 1791.

Rottboell, C.F. 1773. *Descriptionum et Iconum rariores.* Koebenhavn.

Roxburgh, W. 1820-1824. *Flora Indica* ed. 1; ed. 2. 1832. Serampore.

Roy, G.P. 1979. The Genus *Eragrostis* P. Beauv. in Rajasthan. *Bull. Bot. Surv. India* 18 (1-4): 102-108.

Roy, G.P. 1984. *Grasses of Madhya Pradesh*. BSI, Howrah.

Sabnis, S.D. 1962. Cyperaceae of Gujarat. *Bull. Bot. Surv. India* 4: 193-197, 199-201.

Sabnis, S.D. and A.R. Chavan. 1967. Cyperaceae of Dangs forests, Gujarat State. *Indian Forester* 193: 190-192.

Sabnis, S.D. and J.V. Joshi. 1979. Notes on two interesting sedges from Gujarat State, India. Ibid. 76: 210-211.

Sabnis, S.D. and S.J. Bedi. 1971. The genus *Fuirena* (Cyperaceae) in Gujarat. *J. Bombay Nat. Hist. Soc.* 68: 857-858.

Sahni, K.C., K.M. Vaid and H.B. Naithani. 1972. Additions to the Cyperaceae of Madhya Pradesh. *Indian Forester* 98: 192-194.

Sampath, S. 1963. The genus *Oryza*: its taxonomy and species relationship. *Oryza* 1: 1-29.

Santapau, H. 1951. A review of Mooney's Supplement to the Botany of Bihar and Orissa. *J. Bombay Nat. Hist. Soc.* 49: 768-770.

Santapau, H. 1958. *History of Botanical Researches in India, Burma and Ceylon*, pt. II. Bangalore.

Sanyal, A. 1957. Additional notes on the Botany of Bihar and Orissa by H.H. Haines and its Supplement by Dr. Mooney. *Indian For.* 83: 230-235.

Savile, D.B.O. 1979. Fungi as aids in the higher plant classification. *Bot. Rev.* 45: 377-503.

Saxena, H.O. 1973. Contributions to the flora of Madhya Pradesh–II (Interesting records of Cyperaceae). *Indian Forester* 99: 505-507.

Sharma, A. 1985. The genus *Panicum* L. in Punjab (India). *J.Econ. Taxon. Bot.* 6 (1): 103-109.

Sharma, B.D. and M. Sanjappa (Ed.). 1993. *Flora of India*. Vol. 3. Botanical Survey of India, Calcutta.

Sharma, B.D. and N.P. Balakrishnan (Ed.). 1993. *Flora of India*. Vol. 2. Botanical Survey of India, Calcutta.

Sharma, B.D., N.P. Balakrishnan, R.R. Rao and P.K. Hajra (Ed.). 1993. *Flora of India*. Vol. I. Botanical Survey of India, Calcutta.

Shukla, U. 1996. *The Grasses of North-eastern India*. Jodhpur.

Shukla, U. and S.K. Jain. 1978. On the nomenclature of some Indian grasses. *Bull. Bot. Surv. Ind.* 20 (1-4): 64-68.

Simon, B.K. 1972. A revision of the genus *Sacciolepis* (Gramineae) in the 'Flora Zambesiaca' area. *Kew Bull.* 27 (3): 387-406.

Singh, Ayodhya, G.K. Pandey and K.K. Mishra. 1998. *Floristic Composition and Vegetational Profile of Aquatic Angiosperms of Hazaribag and Adjacent Area.* Ph.D. Thesis, Vinoba Bhave University, Hazaribag.

Singh, G. 1986. Some recent additions and name changes concerning Indian Poaceae. *J. Econ. Tax. Bot.* 8 (2): 493-503.

Singh, N.P., U.R. Deshpande and R.S. Raghavan. 1979. Poaceae of Karnataka State. *Bull Bot. Surv. Ind.* 18 (1-4): 109-143. (pub. 1976).

Singh, N.P., V. Mudgal, K.K. Khanna, S.C. Srivastava, A.K. Sahoo, S. Bandopadhyay, N. Aziz, M. Das, R.P. Bhattacharya and P.K. Hajra. 2001. *Flora of Bihar Analysis.* BSI.

Sivagnanam, L. 1960. Extra-axillary digitate spike in *Eleusine coracana* Gaertn. *Madras Agric. J.* 47: 329.

Sivagnanam, L. 1961. Rachis branching in the "Finger millet" *Eleusine coracana* Gaertn. (Ragi). *Madras Agric. J.* 48: 221-222.

Skerman, P.J. and F. Riveros. 1990. *Tropical grasses.* Scientific Publishers, Jodhpur.

Sreekumar, P.V. and V.J. Nair. 1991. *Flora of Kerala-Grasses.* BSI, Calcutta.

Srivastava, J.G. 1959. Recent trends in the flora of Bihar states. *J. Indian Bot. Soc.* 38: 186-194.

Srivastava, J.G. 1964. Some tropical American and African Weeds that have invaded the state of Bihar. *J. Indian Bot. Soc.* 43: 102-112.

Srivastava, S.K. and B.K. Sinha. 2000. Genus *Ischaemum* L. (Poaceae) in India. In Gupta, B.K., (ed.). *Higher Plants of Indian Subcontinent,* Vol. 9. Bishen Singh Mahendra Pal Singh, Dehra Dun.

Stafleu, F.A. and E.A. Mennega. 1992-2000. *Taxonomic Literature: Supplement.* Koeltz Scientific Books, Koenigstein, Germany.

Steudel, E.G. 1854. *Synopsis plantarum glumacerum* Vol. 2. *Synopsis plantarum Cyperacearum,* fasc. 7: 1-80. et fasc. 8-10: 81-348. 1855.

Sur, P.R. 1988. A taxonomic revision of the genus *Sehima* Forssk. (Poaceae) in India. *J. Econ. Taxon. Bot.*12 (1): 71-79.

Sur, P.R. 2001. A revision of the genus *Ischaemum* Linn. (Poaceae) in Inida. *J. Econ. Taxon. Bot.* 25 (2): 407-437.

Takhtajan, Arman. 1969. *Flowering Plants: Origin and Dispersal.* Oliver and Boyd, Edinburgh.

Tournefort, J.P. de. 1719. *Institutiones rei herbariae* ed. 3. Paris.

Trivedi, G.N., H.A. Molla and D.C. Pal. 1985. Some uses of plants from tribal areas of Chotanagpur, Bihar. *Nagarjun* 29 (2): 15-18.

Tutin, T.G., V.H. Heywood, N.A. Burges, D.H. Valentine, S.M. Walters and D.A. Webb (eds.). 1980. *Flora Europaea.* Vol. 5. Poaceae. Cambridge University Press.

Uniyal, B.P. 1986. *Deyeuxia borii* Bhattacharya and Jain – a superfluous name. *J. Econ. Tax. Bot.* 8: 235.

Uppuluri, M.R. 1975 (1972). Taxonomy of the genus *Apluda* L. *Bull. Bot. Surv. India.* 14: 149-151.

Vahl, M. 1805/1806. *Enumeration Plantarum* Vol. 2. Copenhagon.

Veldkamp, J.F. 1999. A revision of *Chrysopogon* Trin. and *Vetiveria* Bory (Gramineae) in Thailand and Malesia with notes on some other species from Africa and Australia. *Austrobaileya* 5: 503-533.

Veldkamp, J.F. 2001. Notes on some species of *Chloris* (Poaceae) described for the Phillippines by P. Durand. *Taxon* 50: 845-852.

Watson, L. and M.J. Dallwitz. 1988. *Grass Genera of the World*: Illustrations of characters, descriptions, classification, interactive identification, information retrieval. Australian National University, Canberra.

Watt, G. 1889-1896. *The Dictionary of Economic Products of India.* Vol. 6. Calcutta.

Willdenow, C.L. 1797-1830. Caroli a Linne *Species Plantarum* ed. 4. Vol. 1-16.

Wood, J.J. 1902. Plants of Chutianagpur, including Jashpur and Sirguja. *Rec. Bot. Surv. India* 2 (1): 1-170.

Yabuno, T. 1966. Biosystematic study of the genus *Echinochloa. Jap. J. Bot.* 19: 277-323.

Yadav, S.R. and M.M. Sardesai. 2002. *Flora of Kolhapur District.* Shivaji Univ., Kolhapur.

Youngner, V.B. and C.M. Mckell. 1972. *The Biology and Utilization of Grasses.* New York and London.

Graph - 1: Frequency of the Sedges

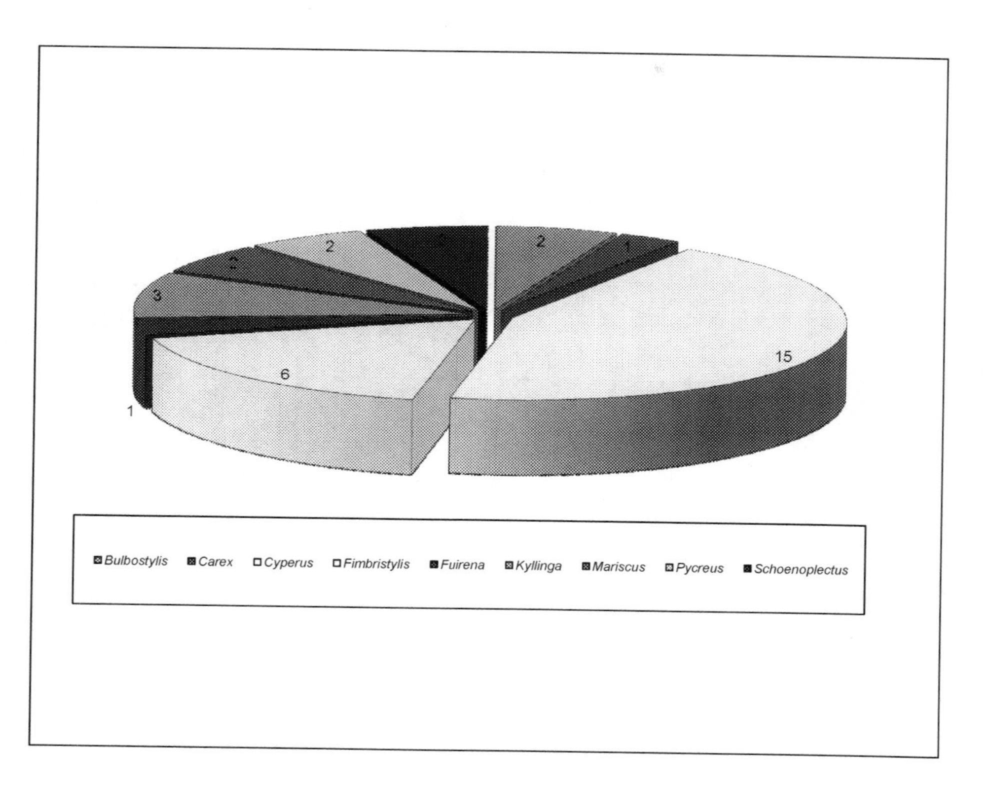

	G				P				S				H			
	Genera	**spp.**	**ssp.**	**var.**	**Genera**	**spp.**	**ssp.**	**var.**	**Genera**	**spp.**	**ssp.**	**var.**	**Genera**	**spp.**	**ssp.**	**var.**
Anosporum									1	1						
Bulbostylis	1	2			1	1			1	3			1	3		
Carex	1	1			1	1			1	10		3	1	6		
Cyperus	1	15	1	1	1	15			1	35	3	6	1	32		2
Eleocharis					1	1			1	8						
Eriophorum									1	1						
Fimbristylis	1	6			1	10	1		1	24		2	1	23		2
Fuirena	1	1			1	1			1	2			1	2		
Hypolytrum									1	1						
Indocourtoisia									1	1						
Kyllinga	1	3			1	3			1	6			1	4		
Lipocarpha									1	2						
Mariscus	1	2			1	2			1	7			1	6		
Pycreus	1	2			1	2			1	7		7	1	7		
Rhynochospora									1	3			1	3		
Rikliella					1	1			1	1						
Schoenoplectus	1	2	1		1	3			1	11						
Scirpus									1	3			1	8		
Scleria					1	1			1	11		2	1	6		
Sorostachys									1	2						

Graph - 2: Comparative account of the genera, species, sub-species and varieties of the Sedges of Gautam Buddha Wildlife Sanctuary (G), Flora of Hazaribagh District (P), Flora of Bihar (S) and Botany of Bihar and Orissa (H)

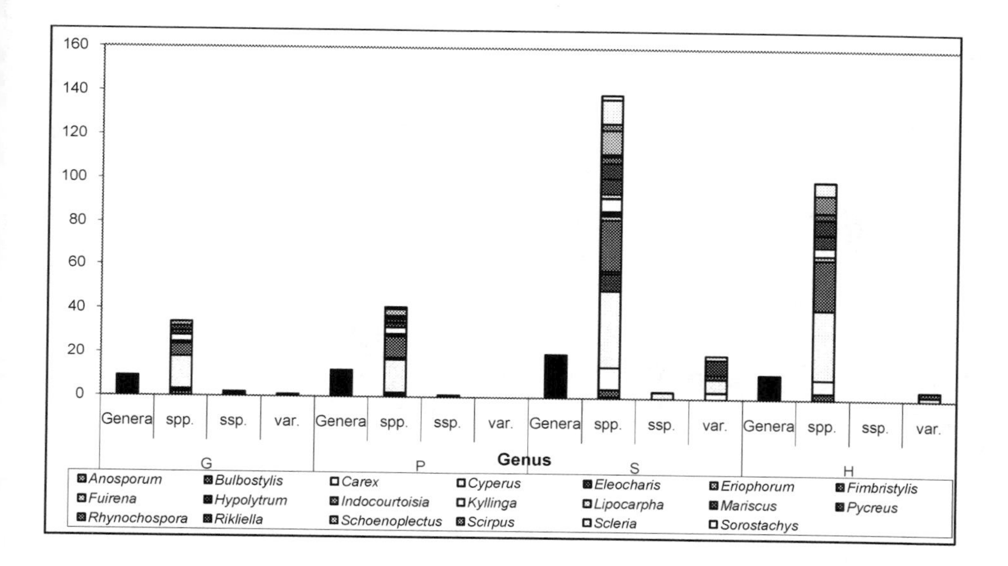

Graph - 3: Frequency of the Grasses

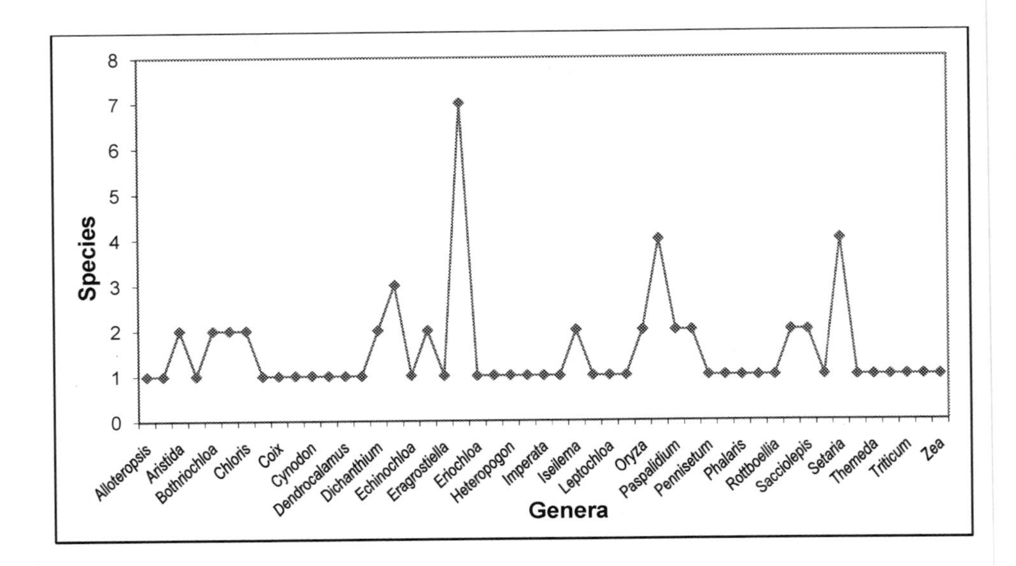

Table - 5: Synopsis of genera, species, sub species/varieties of the Sedges

Sl. No.	Genera	Species	Sub species / variety	Total
1.	*Bulbostylis* Kunth	2		2
2.	*Carex* L.	1		1
3.	*Cyperus* L.	15	1 / 1	17
4.	*Fimbristylis* Vahl	6		6
5.	*Fuirena* Rottb.	1		1
6.	*Kyllinga* Rottb.	3		3
7.	*Mariscus* Vahl	2		2
8.	*Pycreus* Beauv.	2		2
9.	*Schoenoplectus* (Reichb.) Palla	2	1	3
	Total	**34**	**3**	**37**

Table - 6: Frequency of the Sedges

Sl. No.	Genera	Species
1.	*Bulbostylis* Kunth	2
2.	*Carex* L.	1
3.	*Cyperus* L.	15
4.	*Fimbristylis* Vahl	6
5.	*Fuirena* Rottb.	1
6.	*Kyllinga* Rottb.	3
7.	*Mariscus* Vahl	2
8.	*Pycreus* Beauv.	2
9.	*Schoenoplectus* (Reichb.) Palla	2
	Total	**34**

Table - 7: Comparative account of the genera, species, sub-species, and varieties of the Sedges of Gautam Buddha Wild Life Sanctuary (G), Flora of Hazaribagh District (P), Flora of Bihar (S) and Botany of Bihar and Orissa (H)

Sl. No.	Genus	G				P				S				H			
		Genera	spp.	ssp.	var.	Genera	spp.	ssp.	var.	Genera	spp.	ssp.	var.	Genera	spp.	ssp.	var.
1.	*Anosporum*									1	1						
2.	*Bulbostylis*	1	2			1	1			1	3			1	3		
3.	*Carex*	1	1			1	1			1	10		3	1	6		
4.	*Cyperus*	1	15	1	1	1	15			1	35	3	6	1	32		2
5.	*Eleocharis*					1	1			1	8						
6.	*Eriophorum*									1	1						
7.	*Fimbristylis*	1	6			1	10	1		1	24		2	1	23		2
8.	*Fuirena*	1	1			1	1			1	2			1	2		
9.	*Hypolytrum*									1	1						
10.	*Indocourtoisia*									1	1						
11.	*Kyllinga*	1	3			1	3			1	6			1	4		
12.	*Lipocarpha*									1	2						
13.	*Mariscus*	1	2			1	2			1	7			1	6		
14.	*Pycreus*	1	2			1	2			1	7		7	1	7		
15.	*Rhynochospora*									1	3			1	3		
16.	*Rikliella*					1	1			1	1						
17.	*Schoenoplectus*	1	2	1		1	3			1	11						
18.	*Scirpus*									1	3			1	8		
19.	*Scleria*					1	1			1	11		2	1	6		
20.	*Sorostachys*									1	2						
	Total	**9**	**34**	**2**	**1**	**12**	**41**	**1**		**20**	**139**	**3**	**20**	**11**	**100**		**4**

Table - 8: Details of hitherto unreported Sedges of Gautam Buddha Wild Life Sanctuary

Sl. No.	Sedges reported here and not reported by Paria and Chattopadhyay (2005) in Fl. of Hazaribagh District, BSI	Sedges reported by H. H. Haines in Botany of Bihar and Orissa (1922)		Sedges not reported by H. H. Haines in Botany of Bihar and Orissa (1922)	Sedges reported by Singh et al (2001) in Fl. of Bihar, BSI		Sedges not reported by Sigh et al. (2001) Fl. of Bihar
		From Hazaribag	Outside Hazaribag		From Hazaribag	Outside Hazaribag	
1.	*Bulbostylis densa* (Wallich) Hand.		✓		✓		
2.	*Cyperus bulbosus* Vahl			✓	✓	✓	
3.	*C. castaneus* Willd.	✓	✓		✓	✓	
4.	*C. dubius* Rottb.		✓				✓
5.	*C. nutans* Vahl		✓		✓	✓	
6.	*C. pangorei* Rottb.	✓	✓		✓	✓	
7.	*C. pygmaeus* Rottb.		✓			✓	
8.	*Fimbristylis alboviridis* Clarke			✓			✓
9.	*F. littoralis* Gaud.			✓		✓	
10.	*F. polytrichoides* (Retz.) R. Br.			✓			✓
11.	*Mariscus compactus* (Retz.) Boldingh		✓			✓	
12.	*M. paniceus* (Retz.) Vahl		✓		✓	✓	

Table - 9: Flowering calendar of the Sedges of Gautam Buddha Wild Life Sanctuary, Hazaribag

Sl. No.	Name of the Plant	Place of Occurrence	Period of Occurrence	Period of flowering	Frequency of occurrence
1.	*Bulbostylis barbata* (Rottb.) Kunth ex Clarke	Hathia baba, Danua	Annual	July – Oct.	Common
2.	*Bulbostylis densa* (Wallich) Hand.	Chordaha, Danua	Annual	Sept. – Dec.	Common
3.	*Carex cruciata* Wahlenb.	Sanjha, Garmorwa	Perennial	July – Aug.	Uncommon
4.	*Cyperus bulbosus* Vahl.	Bukar, Mohane	Perennial	July – Oct.	Uncommon
5.	*Cyperus castaneus* Willd.	Asnachuan	Annual	Aug. – Dec.	Common
6.	*Cyperus difformis* L.	Asnachuan, Chamargadda	Annual	Aug. – Jan.	Frequent
7.	*Cyperus distans* L.	Danua	Perennial	Aug. – Nov.	Common
8.	*Cyperus dubius* Rottb.	Dhoria, Kabilas	Perennial	June – Nov.	Uncommon
9.	*Cyperus halpan* L.	Khairtand, Siarkoni	Perennial	June – Oct.	Common
10.	*Cyperus iria* L.	Garmorwa, Sanjha, Siarkoni	Annual	Aug. – Dec.	Common
11.	*Cyperus nutans* Vahl.	Chordaha, Danua	Perennial	July – Dec.	Uncommon

Sl. No.	Name of the Plant	Place of Occurrence	Period of Occurrence	Period of flowering	Frequency of occurrence
12.	*Cyperus pangorei* Rottb.	Silodhar, Murtiakalan	Perennial	Aug. – Nov.	Uncommon
13.	*Cyperus platystylis* R. Br.	Bukar, Khairtand	Perennial	May – Nov.	Common
14.	*Cyperus polystachyos* Rottb.	Kabilas, Mainukhar	perennial	Aug. – Dec.	Uncommon
15.	*Cyperus procerus* Rottb.	Kabilas, Silodhar	Perennial	Sept. – Nov.	Common
16.	*Cyperus pygmaeus* Rottb.	Silodhar, Dhoria	Annual	Aug. – Nov.	Uncommon
17.	*Cyperus rotundus* L.	Asnachuan, Bukar	Perennial	July – Dec.	Frequent
18.	*Cyperus tenuispica* Steud.	Pathalgarwa, Khairtanr	Annual	May – Dec.	Common
19.	*Fimbristylis alboviridis* C. B. Clark	Chordaha, Danua	Annual	Sept. – Nov.	Common
20.	*Fimbristylis argentea* (Rottb.) Vahl.	Mohane tand, Chordaha	Annual	Nov. – Jan.	Common
21.	*Fimbristylis littoralis* Gaud	Hathia baba, Kabilas	Perennial	Sept. - April	Common
22.	*Fimbristylis miliacea* (L.) Vahl.	Duragara, Mainukhar	Annual	Aug.- Dec.	Common

Sl. No.	Name of the Plant	Place of Occurrence	Period of Occurrence	Period of flowering	Frequency of occurrence
23.	*Fimbristylis polytrichoides* (Retz.) R. Br.	Chamargadda, Garmorwa	Perennial	July – Sep.	Uncommon
24.	*Fimbristylis tetragona* R. Br.	Mohane tand, Garmorwa	Perennial	Aug. – Nov.	Common
25.	*Fuirena ciliaris* (L.) Roxb.	Duragara, Pathalgara	Annual	Oct. – Jan.	Common
26.	*Kyllinga brevifolia* Rottb.	Bukar, Sikda	Perennial	Throughout year especially from June to November	Common
27.	*Kyllinga bulbosa* Beauv.	Sanjha, Siarkoni	Perennial	Aug. – Feb.	Common
28.	*Kyllinga nemoralis* (J.R. and G. Forster) Dandy ex Hutchinson and Dalziel	Muria, Dhoria	Perennial	July – Feb.	Common
29.	*Mariscus compactus* (Retz.) Boldingh	Chordaha, Danua	Perennial	Sept. – Jan.	Common
30.	*Mariscus paniceus* (Rotttb.) Vahl.	Sanjha, Siarkoni, Mainukhar	Perennial	July – Sept.	Common
31.	*Pycreus flavidus* (Retz.) Koyama	Khairtanr, Sanjha	Perennial	Feb. - May	Common

Sl. No.	Name of the Plant	Place of Occurrence	Period of Occurrence	Period of flowering	Frequency of occurrence
32.	*Pycreus pumilus* (L.) Nees ex Clarke	Muria, Silodhar	Annual	July – Nov.	Common
33.	*Schoenoplectus articulatus* (L.) Palla	Mohane tand, Muria	Perennial	Nov. – Jan.	Common
34.	*Schoenoplectus supinus* (L.) Palla	Murtia Kalan, Silodhar	Annual	Sept. – Feb.	Common

Table - 10: Flowering calendar and growth pattern of Grasses found in Gautam Buddha Wild Life Sanctuary, Hazaribag

Sl. No.	Name of the Plant	Place of occurrence	Period of occurrence	Period of flowering	Frequency of occurrence
1.	*Alloteropsis cimicina* (L.)	Ahri, Kathodumar	Annual	July - Oct.	Common
2.	*Apluda mutica* L.	Mainukhar, Sanjha	Perennial	Sept. - Nov.	Uncommon
3.	*Aristida adscensionis* L.	Bukar, Pathalgarwa	Annual	Aug. - March	Common
4.	*Aristida setacea* Retz.	Silodhar, Morainia	Perennial	Aug. - Dec.	Common
5.	*Bambusa arundinacea* (Retz.) Willd.	Hathia baba, Garmorwa	Perennial	Once in Life	Common
6.	*Bothriochloa bladhii* (Retz.) S.T. Blake	Danua, Chordaha	Perennial	Aug. - Jan.	Common
7.	*Bothriochloa pertusa* (L.) A. Camus	Chordaha, Garmorwa, Pathalgara	Perennial	Through out the year if moisture is available	Frequent
8.	*Brachiaria ramosa* (L.) Stapf	Sanjha, Khairtanr	Annual	June - Dec.	Frequent
9.	*Brachiaria reptans* (L.) Gardner and Hubbard	Dhoria, Muria	Annual	Aug. - Oct.	Common
10.	*Chloris barbata* Sw.	Kathodumar, Kabilas	Perennial	Aug. - Jan.	Common
11.	*Chloris dolichostachya* Lagasca	Murtiakalan, Bukar	Perennial	Sept. - Dec.	Uncommon

Sl. No.	Name of the Plant	Place of occurrence	Period of occurrence	Period of flowering	Frequency of occurrence
12.	*Chrysopogon aciculatus* (Retz.) Trin.	Danua, Sanjha	Perennial	Almost through out the year but mainly during May to September	Frequent
13.	*Coix lachryma-jobi* L.	Ahri, Chordaha	Perennial	Aug. – Jan.	Uncommon
14.	*Cymbopogon martinii* (Roxb.) Wats.	Silodhar	Perennial	Sept. – Dec.	Uncommon
15.	*Cynodon dactylon* (L.) Pers.	Sikda, Mainukhar	Perennial	All round the year	Frequent
16.	*Dactyloctenium aegyptium* (L.) Beauv.	Morainia, Chordaha, Garmorwa	Annual	June – Oct.	Frequent
17.	*Dendrocalamus strictus* (Roxb.) Nees	Garmorwa	Perennial	Oct. – Nov.	Common
18.	*Desmostachya bipinnata* (L.) Stapf	Ahri	Perennial	June – Oct.	Common
19.	*Dichanthium annulatum* (Forssk.) Stapf	Muria, silodhar	Perennial	Nov. - June	Frequent
20.	*Dichanthium ischaemum* (L.) Roberty	Morainia, Asnachuan	Perennial	Aug. – Dec.	Uncommon
21.	*Digitaria bicornis* (Lam.) Roem and Schult. ex Loud.	Danua, Pathalgarwa, Garmorwa			
22.	*Digitaria ciliaris* (Retz.) Koeler	Dhoria, Murtiakalan	annual	May – Nov.	Frequent

Sl. No.	Name of the Plant	Place of occurrence	Period of occurrence	Period of flowering	Frequency of occurrence
23.	*Digitaria longiflora* (Retz.) Pers.	Sanjha, Kathodumar	Annual	July – Dec.	Uncommon
24.	*Echinochloa colona* (L.) Link	Duragara, Asnachuan	Annual	July – Oct.	Frequent
25.	*Eleusine coracana* (L.) Gaertn.	Morainia, Silodhar	Annual	Sept. – Dec.	Common
26.	*Eleusine indica* (L.) Gaertn.	Bukar, Mainukhar	Annual	Aug. – Nov.	Frequent
27.	*Eragrostiella bifaria* (Vahl) Bor	Chordaha, Pathalgarwa	Perennial	Sept. – Dec.	Uncommon
28.	*Eragrostis atrovirens* (Desf.) Trin. ex Steud.	Moraina, Muria	Perennial	Aug. – Nov.	Common
29.	*Eragrostis ciliata* (Roxb.) Nees	Pathalgara, Duragara	Perennial	Oct. - Dec.	Uncommon
30.	*Eragrostis gangetica* (Roxb.) Steud.	Dhoria, Kabilas	Annual	Sept. – Jan.	Common
31.	*Eragrostis nutans* (Retz.) Nees ex Steud.	Chordaha, Danua	Perennial	Sept. – Jan.	Common
32.	*Erogrostis tenella* (L.) P. Beauv. ex Roem. and Schult.	Ahri, Sanjha, Garmorwa	Annual	Aug. – Feb.	Frequent
33.	*Eragrostis tremula* (Lam.) Hochst. ex. Steud.	Kathodumar, Bukar	Annual	Oct. – Dec.	Common
34.	*Eragrostis unioloides* (Retz.) Nees ex Steud.	Pathalgarwa, Khairtanr	Annual	Oct. – Jan	Frequent.

Sl. No.	Name of the Plant	Place of occurrence	Period of occurrence	Period of flowering	Frequency of occurrence
35.	*Eriochloa procera* (Retz.) C.E. Hubb.	Asnachuan, Sikda	Perennial	Aug. – Dec.	Uncommon
36.	*Hackelochloa granularis* (L.) O. Kuntze Tripali, Kangani	Kabilas	Annual	Aug. – Nov.	Uncommon
37.	*Heteropogon contortus* (L.) P. Beauv. ex Roem and Schult.	Murtiakalan, Sikda	Perennial	Sept. – Jan.	Frequent
38.	*Hordeum vulgare* L.	Ahri, Garmorwa	Annual	Jan. - March	Common
39.	*Imperata cylindrica* (L.) Beauv.	Pathalgarwa, Duragara	Perennial	March – June andOct.–Nov.	Common
40.	*Ischaemum indicum* (Houtt.) Merr.	Kathodumar, Silodhar	Perennial	Aug. – Dec.	
41.	*Iseilema anthephoroides* Hack.	Muria, Dhoria	Annual	July – Dec.	Uncommon
42.	*Iseilema prostratum* (L.) Anderss.	Asnachuan, Mainukhar	Perennial	Aug. – Nov.	Common
43.	*Leersia hexandra* Sw.	Kenduadih, Sanjha	Perennial	Sept. – Dec.	Uncommon
44.	*Leptochloa panicea* (Retz.) Ohwi	Chordaha, Pathagara	Annual	June – Oct.	Uncommon
45.	*Oplismenus burmanii* (Retz.) P. Beauv.	Sikda, Asnachuan	Annual	Sept. – Dec.	Frequent

Sl. No.	Name of the Plant	Place of occurrence	Period of occurrence	Period of flowering	Frequency of occurrence
46.	*Oryza rufipogon* Griff.	Danua, Chordaha, Bukar, Pathalgarwa	Annual	Sept. – Nov.	Common
47.	*Oryza sativa* L.	Chordaha, Bukar	Annual	Sept. – Dec.	Frequent
48.	*Panicum paludosum* Roxb.	Muria, Mainukhar	Perennial	Aug. – Dec.	Common
49.	*Panicum psilopodium* Trin.	Ahri, Silodhar	Annual	Aug. – Nov.	Common
50.	*Panicum repens* L.	Sikda, Mainukhar	Perennial	Aug. – Sept.	Frequent
51.	*Panicum trypheron* Schult.	Morainia, Kathodumar	Annual	Aug. – Dec.	Uncommon
52.	*Paspalidium flavidum* (Retz.) A. Camus	Bukar, Asnachuan	Perennial	Aug. – Nov.	Frequent
53.	*Paspalidium geminatum* (Forssk.) Stapf	Sanjha, Pathalgara	Perennial	July – Dec.	Uncommon
54.	*Paspalum distichum* L.	Kabilas, Silodhar	Perennial	Aug. – Nov.	Uncommon
55.	*Paspalum scrobiculatum* L.	Garmorwa, Duragara	Annual	Aug. – Nov.	Common
56.	*Pennisetum pedicellatum* Trin.	Sikda, Chordaha	Annual	Sept. – Dec.	Common
57.	*Perotis indica* (L.) O. Kuntze	Garmorwa, Pathalgara	Annual	Oct. – Nov.	Common
58.	*Phalaris minor* Retz.	Mainukhar, Ahri	Annual	Feb. – Apr.	Common

Sl. No.	Name of the Plant	Place of occurrence	Period of occurrence	Period of flowering	Frequency of occurrence
59.	*Polypogon monspeliensis* (L.) Desf.	Kathodumar, Sanjha	Annual	Feb. – Apr.	Common
60.	*Rottboellia cochinchinensis* (Lour.) W.D. Clayton	Garmorwa, Sikda	Annual	Sept. – Dec.	Uncommon
61.	*Saccharum bengalense* Retz.	Hathia baba, Danua	Perennial	Oct. – Dec.	Common
62.	*Saccharum officinarum* L.	Pathagarwa Chordaha	Perennial	Feb. - April	Common
63.	*Saccharum spontaneum* L.	Dhoria, Chordaha Kabilas	Perennial	Aug. – Nov.	Frequent
64.	*Sacciolepis interrupta* (Willd.) Stapf	Duragara	perennial	Sept. – Nov.	Uncommon
65.	*Sacciolepis myosuroides* (R. Br.) A. Camus	Silodhar, Murtiakalan	Annual	Aug. – Dec.	Uncommon
66.	*Sehima nervosum* (Rottb.) Stapf	Pathalgara, Chordaha	Perennial	Aug. – Dec.	Uncommon
67.	*Setaria glauca* (L.) P. Beauv	Kenduadih, Sikda	Annual	Aug. – Oct.	Common
68.	*Setaria intermedia* Roem. and Schult.	Chordaha	Annual	July – Oct.	Common
69.	*Setaria pumila* (Poir.) Roem. and Schult	Muria	Annual	Almost all the year round	Common
70.	*Setaria verticillata* (L.) P. Beauv.	Pathalgara, Murtiakalan	Annual	Almost throughout the year	Uncommon

Sl. No.	Name of the Plant	Place of occurrence	Period of occurrence	Period of flowering	Frequency of occurrence
71.	*Sporobolus diander* (Retz.) P. Beauv.	Duragara, Kabilas, Dhoria	Perennial	Thought out the but mainly in rainy season	Frequent
72.	*Themeda quadrivalvis* (L.) O. Kuntze	Bukar, Pathalgarwa	Annual	Sep. – Oct.	Common
73.	*Thysanolaena maxima* (Roxb.) Kuntze	Ahri, Pathalgarwa	Annual	Apr. - June	Common
74.	*Triticum aestivum* L.	Khairtanr, Chordaha	Annual	Jan. - May	Frequent
75.	*Vetiveria zizanioides* (L.) Nash	Chordaha, Sikda	Perennial	Aug. – Dec.	Common
76.	*Zea mays* L.	Danua, Chordaha	Annual	May – Oct	Frequent

Table - 11: Name of Villages/Localities having maximum/ minimum representation of the Sedges and Grasses

Sl. No.	Name of the Localities	Species found
1.	Ahri	+ + + + + + + +
2.	Asnachuan	+ + + + + + + + + +
3.	Bukar	+ + + + + + + + + + + +
4.	Chamargadda	+ +
5.	Chordaha	+ +
6.	Danua	+ + + + + + + + + + + + + +
7.	Dhoria	+ + + + + + + + +
8.	Duragara	+ + + + + + + +
9.	Garmorwa	+ + + + + + + + + + + + + + +
10.	Hathia baba	+ + + +
11.	Kabilas	+ + + + + + + + + +
12.	Kathodumar	+ + + + + + +
13.	Kenduadih	+ +
14.	Khairtand	+ + + + + + +
15.	Mainukhar	+ + + + + + + + + +
16.	Mohane tand	+ + + +
17.	Morainia	+ + + + + + + +
18.	Muria	+ + + + + + + + +
19.	Murtiakalan	+ + + + + + +
20.	Pathalgarwa	+ + + + + + + +
21.	Pathalgara	+ + + + + + + + + + +
22.	Sanjha	+ + + + + + + + + + + + +
23.	Siarkoni	+ + + +
24.	Sikda	+ + + + + + + + + +
25.	Silodhar	+ + + + + + + + + + + + + + +

Note: - '+' = Number of species

Table - 12: Comparison of Dominant Grass genera of Gautam Buddha Wildlife Sanctuary, Hazaribag, Bihar (Jharkhand) and India

Sl. No.	Gautam Buddha Wildlife Sanctuary, Hazaribag	Hazaribag Paria and Chattopadhyay (2005)	Bihar Singh et al. (2001)	India S. Maulik (1997)
1.	*Aristida* (2 spp.)	*Cynodon* (3 spp.)	*Aristida* (7 spp.)	*Arthraxon* (23 spp.)
2.	*Bothriochloa* (2 spp.)	*Dichanthium* (3 spp.)	*Brachiaria* (9 spp.)	*Bothriochloa* (17 spp.)
3.	*Brachiaria* (2 spp.)	*Digitaria* (4 spp.)	*Chrysopogon* (8 spp.)	*Brachiaria* (18 spp.)
4.	*Dichanthium* (2 spp.)	*Echinochloa* (3 spp.)	*Cymbopogon* (11 spp.)	*Cymbopogon* (21 spp., var. 14, ssp. 1)
5.	*Digitaria* (3 spp.)	*Eragrostis* (10 spp.)	*Digitaria* (11 spp., var. 2)	*Digitaria* (27 spp., var. 19, ssp. 3)
6.	*Eragrostis* (7 spp.)	*Oryza* (2 spp.)	*Eragrostis* (21 spp.)	*Eragrostis* (35 spp., var. 7)
7.	*Panicum* (4 spp.)	*Panicum* (5 spp.)	*Panicum* (11 spp.)	*Isachne* (12 spp.)
8.	*Saccharum*(3 spp.)	*Paspalum* (2 spp.)	*Pennisetum* (10 spp.)	*Ischaemum* (43 spp., var. 3)
9.	*Sacciolepis*(2 spp.)	*Saccharum* (2 spp.)	*Setaria* (10 spp.)	*Panicum* (31 spp., var. 2)
10.	*Setaria* (4 spp.)	*Setaria* (4 spp.)	*Themeda* (9 spp. var. 2)	*Themeda* (19 spp., var. 2)

Table - 13: Details of hitherto unreported Grasses of Gautam Buddha Wildlife Sanctuary

Sl. No.	Grasses reported here and not reported by Paria and Chattopadhyay (2005) in Fl. of Hazaribagh District, BSI	Grasses reported by H. H. Haines(1924) in Botany of Bihar and Orissa		Grasses not reported by H. H. Haines (1924) in Botany of Bihar and Orissa	Grasses reported by Singh et al (2001) in Fl. of Bihar, BSI		Grasses not reported by Sigh et al. (2001) in Fl. of Bihar, BSI
		From Hazaribag	Outside Hazaribag		From Hazaribag	Outside Hazaribag	
1.	*Aristida adscensionis* L.	✓			✓		
2.	*Bothriochloa pertusa* (L.) A. Camus			✓	✓		
3.	*Brachiaria reptans* (L.) Gardner & Hubbard		✓			✓	
4.	*Chloris dolichostachya* Lagasca		✓		✓		
5.	*Desmostachya bipinnata* (L.) Stapf		✓			✓	
6.	*Digitaria bicornis* (Lam.) Roem. & Schult. ex Loud.			✓		✓	
7.	*Iseilema anthephoroides* Hack.		✓			✓	
8.	*Paspalidium geminatum* (Forssk.) Stapf		✓			✓	
9.	*Sacciolepis interrupta* (Willd.) Stapf		✓			✓	
10.	*Sacciolepis myosuroides* (R. Br.) A. Camus	✓			✓		
11.	*Sehima nervosum* (Rottb.) Stapf	✓			✓		
12.	*Setaria pumila* (Poir) Roem. & Schult.			✓		✓	
13.	*Vetiveria zizanioides* (L.) Nash	✓			✓		

Table - 14: Grasses of Ecological Significance growing in Gautam Buddha Wildlife Sanctuary

Sl. No.	Drought resistant grass	Soil binders of river and stream beds	Fighting soil erosion on hilly tracks	Grasses colonising mine dunes
1.	*Aristida adscensionis* L.	*Cynodon dactylon* (L.) Pers.	*Alloteropsis cimicina* (L.) Stapf	*Alloteropsis cimicina* (L.) Stapf
2.	*Chloris barbata* Sw.	*Demostachya bipinnata* (L.) Stapf	*Cymbopogon martinii* (Roxb.) Wats.	*Aristida setacea* Retz.
3.	*Cynodon dactylon* (L.) Pers.	*Eleusine indica* (L.) Gaertn.	*Cynodon dactylon* (L.) Pers.	*Digitaria ciliaris* (Retz.) Koeler
4.	*Desmostachya bipinnata* (L.) Stapf	*Eragrostiella bifaria* (Vahl) Bor	*Dactyloctenium aegyptium* (L.) Beauv.	*Digitaria longifolia* (Retz.) Pers.
5.	*Echinochloa colona* (L.) Link	*Eragrostis tenella* (L.) P. Beauv.	*Digitaria bicornis* (Lam.) Roem. & Schult. ex Loud.	*Saccharum bengalense* Retz.
6.	*Eleusine indica* (L.) Gaertn.	*Panicum paludosum* Roxb.	*Eleusine indica* (L.) Gaertn.	*Saccharum spontaneum* L.
		Panicum psilopodium Trin.	*Eragrostis tenella* (L.) P. Beauv.	
		Paspalidium flavidum (Retz.) A. Camus	*Eragrostis tremula* (Lam.) Hochst. ex Steud.	
		Paspalum distichum L.		

Table - 15: List of Grasses which are cultivated by the people residing in or around Gautam Buddha Wildlife Sanctuary, Hazaribag

Sl. No.	Name of the Plants
1.	*Bambusa arundinacea* (Retz.) Willd.
2.	*Dendrocalamus strictus* (Roxb.) Nees.
3.	*Eleusine coracana* (L.) Gaertn.
4.	*Hordeum vulgare* L.
5.	*Oryza sativa* L.
6.	*Saccharum Officinarum* L.
7.	*Triticum aestivum* L.
8.	*Zea mays* L.

Table - 16: Grasses which are found in dry condition

Sl. No.	Name of the Plants
1.	*Bambusa arundinacea* (Retz.) Willd.
2.	*Bothriochloa pertusa* (L.) A. Camus
3.	*Chrysopogon aciculatus* (Retz.) Trin.
4.	*Dendrocalamus strictus* (Roxb.) Nees
5.	*Desmostachya bipinnata* (L.) stapf
6.	*Dichanthium annulatum* (Forssk.) stapf
7.	*Eragrostis nutans* (Retz.) Nees ex steud.
8.	*Saccharum bengalense* Retz.
9.	*Saccharum spontaneum* L.

Table - 17: Grasses which can be planted after forest fire

Sl. No.	Name of the Plants
1.	*Cynodon dactylon* (L.) Pers.
2.	*Chrysopogon aciculatus* (Retz.) Trin.
3.	*Eragrostis nutans* (Retz.) Nees ex Steud.
4.	*Saccharum bengalense* Retz.
5.	*Saccharum spontaneum* L.

Table - 18: List of the plants used for forestation

Sl. No.	Name of the Plants
1.	*Bambusa arundinacea* (Retz.) Wild.
2.	*Dendrocalamus strictus* (Roxb.) Nees
3.	*Saccharum bengalense* Retz.

Table - 19: Showing Grasses growing on walls and their nature of occurrence

Sl. No.	Frequent	Less frequent	Occasional
1.	*Chloris barbata* Sw.		
2.		*Apluda mutica* L.	
3.	*Digitaria ciliaris* (Retz.) Koeler		
4.			*Bothriochloa bladhii* (Retz.) S.T. Blake
5.		*Aristida adscensionis* L.	
6.		*Brachiaria ramosa* (L.) Stapf	
7.			*Dactyloctenium aegyptium* (L.) Beauv.

8.		*Setaria verticillata* (L.) P. Beauv.	
9.			*Eleusine indica* (L.) Gaertn.
10.	*Heteropogon contortus* (L.) P. Beauv. ex Roem & Schult.		
11.			*Sporobolus diander* (Retz.) P. Beauv.

Table - 20: Showing Grasses found in the crevices of rocks and their nature of occurrence

Sl. No.	Frequent	Less frequent	Occasional
1.	*Apluda* L.		
2.		*Chloris* Sw.	
3.			*Cymbopogon* Spreng.
4.	*Aristida* L.		
5.		*Eragrostiella* Bor	
6.	*Dendrocalamus* Nees		
7.		*Heteropogon* Pers.	
8.		*Sehima* Forssk.	
9.		*Sporobolus* R. Br.	
10.			*Setaria* Beauv.

Table - 21: Grasses which grow in waste lands

Sl. No.	Grasses	Gregarious	Frequent	Less frequent
1.	*Alloteropsis* C. Presl.			✓
2.	*Aristida* L.		✓	
3.	*Bothriochloa* O. Kuntze	✓		
4.	*Chloris* Sw.			✓
5.	*Dactyloctenium* Willd.	✓		
6.	*Desmostachya* Stapf		✓	
7.	*Digitaria* Rich.	✓		
8.	*Eleusine* Gaertn.	✓		
9.	*Eragrostis* Beauv.	✓		
10.	*Iseilema* Anders.			✓
11.	*Panicum* L.		✓	
12.	*Paspalum* L.		✓	
13.	*Setaria* Beauv.		✓	

Table - 22: Grasses used in Hindu Rituals

Sl. No.	Grasses	Rituals	Parts used	Item
1.	*Aristida setacea* Retz.	Used at rituals witch craft	Culm + fruit	Wild
2.	*Cynodon dactylon* (L.) Pers.	Marriage ceremony, various Pujas, Yagyo Pavit	Culm part + Leaves	Wild + Cultivated
3.	*Desmostachya bipinnata* (L.) Stapf	Religious ceremony, Sarswati Puja	Leaves are used	Wild
4.	*Hordeum vulgare* L.	Religious ceremony, Durga Puja	Grain + off shoots	Cultivated
5.	*Imperata cylindrica* (L.) P. Beauv.	Religious ceremony	Leaves are used	Wild
6.	*Saccharum officinarum* L.	Chhath, Dev Utthan Puja	Stem/Culm	Cultivated
7.	*Triticum aestivum* L.	Chhath, Religious ceremony, Holi	Grains + spikes	Cultivated

Table - 23: Frequency of the Grasses

Sl. No.	Genera	Species
1.	*Alloteropsis* (L.) Stapf	1
2.	*Apluda* L.	1
3.	*Aristida* L.	2
4.	*Bambusa* Schreb.	1
5.	*Bothriochloa* O. Kuntze	2
6.	*Brachiaria* Griseb.	2
7.	*Chloris* Sw.	2
8.	*Chrysopogon* Trin.	1
9.	*Coix* L.	1
10.	*Cymbopogon* Spreng.	1
11.	*Cynodon* Rich. ex Pers.	1
12.	*Dactyloctenium* Willd.	1
13.	*Dendrocalamus* Nees	1
14.	*Desmostachya* Stapf	1
15.	*Dichanthium* Willemet	2
16.	*Digitaria* Rich.	3
17.	*Echinochloa* P. Beauv.	1
18.	*Eleusine* Gaertn.	2
19.	*Eragrostiella* Bor.	1
20.	*Eragrostis* Beauv.	7
21.	*Eriochloa* H.B.K.	1
22.	*Hackelochloa* O. Kuntze	1
23.	*Heteropogon* Pers.	1
24.	*Hordeum* L.	1
25.	*Imperata* Cyrill.	1

Sl. No.	Genera	Species
26.	*Ischaemum* L.	1
27.	*Iseilema* Anders.	2
28.	*Leersia* Sw.	1
29.	*Leptochloa* Beauv.	1
30.	*Oplismenus* Beauv.	1
31.	*Oryza* L.	2
32.	*Panicum* L.	4
33.	*Paspalidium* Stapf	2
34.	*Paspalum* L.	2
35.	*Pennisetum* Rich.	1
36.	*Perotis* Ait.	1
37.	*Phalaris* L.	1
38.	*Polypogon* Desf.	1
39.	*Rottboellia* L. f.	1
40.	*Saccharum* L.	2
41.	*Sacciolepis* Nash	2
42.	*Sehima* Forssk.	1
43.	*Setaria* Beauv.	4
44.	*Sporobolus* R. Br.	1
45.	*Themeda* Forssk.	1
46.	*Thysanolaena* Nees	1
47.	*Triticum* L.	1
48.	*Vetiveria* Bory	1
49.	*Zea* L.	1
	Total	**75**

Table - 24: Synopsis of genera, species, sub species/variety of Grasses

Sl. No.	Genera	Species	Sub species / variety	Total
1.	*Alloteropsis* (L.) Stapf	1		1
2.	*Apluda* L.	1		1
3.	*Aristida* L.	2		2
4.	*Bambusa* Schreb.	1		1
5.	*Bothriochloa* O. Kuntze	2		2
6.	*Brachiaria* Griseb.	2		2
7.	*Chloris* Sw.	2		2
8.	*Chrysopogon* Trin	1		1
9.	*Coix* L.	1		1
10.	*Cymbopogon* Spreng.	1		1
11.	*Cynodon* Rich. ex Pers.	1		1
12.	*Dactyloctenium* Willd.	1		1
13.	*Dandrocalamus* Nees	1		1
14.	*Desmostachya* Stapf	1		1
15.	*Dichanthium* Willemet	2		2
16.	*Digitaria* Rich.	3		3
17.	*Echinochloa* P. Beauv.	1		1
18.	*Eleusine* Gaertn.	2		2
19.	*Eragrostiella* Bor.	1		1
20.	*Eragrostis* Beauv.	7		7
21.	*Eriochloa* H.B.K.	1		1
22.	*Hockelochloa* O. Kuntze	1		1
23.	*Heteropogon* Pers.	1		1
24.	*Hordeum* L.	1		1
25.	*Imperata* Cyrill.	1	1	2

Sl. No.	Genera	Species	Sub species / variety	Total
26.	*Ischaemum* L.	1		1
27.	*Iseilema* Anders.	2		2
28.	*Leersia* Sw.	1		1
29.	*Leptochloa* Beauv.	1		1
30.	*Oplismenus* Beauv.	1		1
31.	*Oryza* L.	2		2
32.	*Panicum* L.	4		4
33.	*Paspalidium* Stapf	2		2
34.	*Paspalum* L.	2		2
35.	*Pennisetum* Rich.	1		1
36.	*Perotis* Ait.	1		1
37.	*Phalaris* L.	1		1
38.	*Polypogon* Desf.	1		1
39.	*Rottboellia* L. f.	1		1
40.	*Saccharum* L.	2		2
41.	*Sacciolepis* Nash	2		2
42.	*Sehima* Forssk.	1		1
43.	*Setaria* Beauv.	4		4
44.	*Sporobolus* R. Br.	1		1
45.	*Themeda* Forssk.	1		1
46.	*Thysanolaena* Nees	1		1
47.	*Triticum* L.	1		1
48.	*Vetiveria* Bory	1		1
49.	*Zea* L.	1		1
	Total	**75**	**1**	**76**

Table - 25: List of Perennial Sedges and Grasses

Sl. No.	Name of the Plants
1.	*Carex cruciata* Wahlenb.
2.	*Cyperus bulbosus* Vahl
3.	*Cyperus distans* L.
4.	*Cyperus dubius* Rottb.
5.	*Cyperus halpan* L.
6.	*Cyperus nutaus* Vahl
7.	*Cyperus pangorei* Rottb.
8.	*Cyperus platystylis* R. Br.
9.	*Cyperus polystachyos* Rottb.
10.	*Cyperus procerus* Rottb.
11.	*Cyperus rotundus* L.
12.	*Fimbristylis littoralis* Gand.
13.	*Fimbristylis polytrichoides* (Retz.) R. Br.
14.	*Fimbristylis tetragona* R. Br.
15.	*Kyllinga brevifolia* Rottb.
16.	*Kyllinga bulbosa* Beauv.
17.	*Kyllinga nemoralis* (J.R. and G. Forster) Dandy ex Hutchinson and Dalziel
18.	*Mariscus compactus* (Retz.) Boldingh
19.	*Mariscus paniceus* (Rottb.) Vahl.
20.	*Pycreus flavidus* (Retz.) Koyama
21.	*Schoenoplectus articulatus* (L.) Palla
22.	*Apluda mutica* L.
23.	*Aristida setacea* Retz.
24.	*Bambusa arundinacea* (Retz.) Willd.
25.	*Bothriochloa bladhii* (Retz.) S.T. Blake

Sl. No.	Name of the Plants
26.	*Bothriochloa pertusa* (L.) A. Camus
27.	*Chloris barbata* Sw.
28.	*Chloris dolichostachya* Lagasca.
29.	*Chrysopogon aciculatus* (Retz.) Trin.
30.	*Coix lachryma-jobi* L.
31.	*Cymbopogon martinii* (Roxb.) Wats.
32.	*Cynodon dactylon* (L.) Pers.
33.	*Dendrocalamus strictus* (Roxb.) Nees
34.	*Desmostachya bipinnata* (L.) Stapf
35.	*Dichanthium annulatum* (Forssk.) Stapf
36.	*Dichanthium ischaemum* (L.) Roberty
37.	*Eragrostiella bifaria* (Vahl) Bor
38.	*Eragostis atrovireus* (Desf.) Trin. ex Steud.
39.	*Eragostis ciliata* (Roxb.) Nees
40.	*Eragostis nutans* (Retz.) Nees ex Steud.
41.	*Eriochloa procera* (Retz.) C.E. Hubb.
42.	*Heteropogon contortus* (L.) P. Beauv. ex Roem. and Schult.
43.	*Imperata cylindrica* (L.) Beauv.
44.	*Ischaemum indicum* (Houtt.) Merr.
45.	*Iseilema prostratum* (L.) Anderss.
46.	*Leersia hexandra* Sw.
47.	*Panicum paludosum* Roxb.
48.	*Panicum repens* L.
49.	*Paspalidium flavidum* (Retz.) A. Camus
50.	*Paspalidium geminatum* (Forssk.) Stapf.
51.	*Paspalum distichum* L.
52.	*Saccharum bengalense* Retz.

Sl. No.	Name of the Plants
53.	*Saccharum officinarum* L.
54.	*Saccharum spontaneum* L.
55.	*Sacciolepis interrupta* (Willd.) Stapf.
56.	*Sehima nervosum* (Rottb.) Stapf.
57.	*Sporobolus diander* (Retz.) P. Beauv.
58.	*Thysanolaena maxima* (Roxb.) Kuntze
59.	*Vetiveria zizanioides* (L.) Nash

Table - 26: Sedges and Grasses which are found in or near water courses

Sl. No.	Name of the Plants	Plants growing in Water lodged area	Marginal Plants
1.	*Bulbostylis barbata* (Rottb.) Kunth ex Clark		✓
2.	*Bulbostylis densa* (Wallich) Hand.		✓
3.	*Carex cruciata* Wahlenb.		✓
4.	*Cyperus bulbosus* Vahl.	✓	
5.	*Cyperus castaneus* Willd.		✓
6.	*Cyperus difformis* L.	✓	
7.	*Cyperus distans* L.	✓	
8.	*Cyperus dubius* Rottb.	✓	
9.	*Cyperus halpan* L.	✓	
10.	*Cyperus iria* L.	✓	
11.	*Cyperus nutans* Vahl.	✓	
12.	*Cyperus pangorei* Rottb.		✓
13.	*Cyperus platystylis* R. Br.	✓	
14.	*Cyperus polystachyos* Rottb.	✓	
15.	*Cyperus procerus* Rottb.	✓	
16.	*Cyperus pygmaeus* Rottb.	✓	
17.	*Cyperus rotundus* L.		✓
18.	*Cyperus tenuispica* Steud.		✓
19.	*Fimbristylis alboviridis* C.B. Clark		✓
20.	*Fimbristylis argentea* (Rottb.) Vahl.		✓

Sl. No.	Name of the Plants	Plants growing in Water lodged area	Marginal Plants
21.	*Fimbristylis littoralis* Gaud.	✓	
22.	*Fimbristylis miliacea* (L.) Vahl.		✓
23.	*Fimbristylis polytrichoides* (Retz.) R. Br.		✓
24.	*Fimbristylis tetragona* R. Br.		✓
25.	*Fuirena ciliaris* (L.) Roxb.		✓
26.	*Kyllinga brevifolia* Rottb.		✓
27.	*Kyllinga bulbosa* Beauv.		✓
28.	*Kyllinga nemoralis* (J.R. & G. Forster) Dandy ex Hutchinson & Dalziel		✓
29.	*Mariscus compactus* (Retz.) Boldingh.	✓	
30.	*Mariscus paniceus* (Rottb.) Vahl.		✓
31.	*Pycreus flavidus* (Retz.) Koyama		✓
32.	*Pycreus pumilus* (L.) Nees ex Clarke		✓
33.	*Schoenoplectus articulatus* (L.) Palla	✓	
34.	*Schoenoplectus supinus* (L.) Palla	✓	
35.	*Alloteropsis cimicina* (L.) Stapf.		✓
36.	*Apluda mutica* L.		✓
37.	*Aristida adscensionis* L.		✓
38.	*Aristida setacea* Retz.		✓
39.	*Bambusa arundinacea* (Retz.) Willd.		✓
40.	*Bothriochloa bladhii* (Retz.) S.T. Blake		✓
41.	*Bothriochloa pertusa* (L.) A. Camus		✓
42.	*Brachiaria ramosa* (L.) Stapf.		✓

Sl. No.	Name of the Plants	Plants growing in Water lodged area	Marginal Plants
43.	*Brachiaria repens* (L.) Gardner & Hubbard		✓
44.	*Chloris barbata* Sw.		✓
45.	*Chloris dolichostachya* Lagasca		✓
46.	*Chrysopogon aciculatus* (Retz.) Trin		✓
47.	*Coix lachryma-jobi* L.		✓
48.	*Cymbopogon martinii* (Roxb.) Wats.		✓
49.	*Cynodon dactylon* (L.) Pers.		✓
50.	*Dactyloctenium aegyptium* (L.) Beauv.		✓
51.	*Dendrocalamus strictus* (Roxb.) Nees		✓
52.	*Desmostachya bipinnata* (L.) Stapf.		✓
53.	*Dichanthium annulatum* (Forssk.) Stapf.		✓
54.	*Dichanthium ischaemum* (L.) Roberty		✓
55.	*Digitaria bicornis* (Lam.) Roem. & Schult. ex Loud.		✓
56.	*Digitaria ciliaris* (Retz.) Koeler		✓
57.	*Digitaria longiflora* (Retz.) Pers.		✓
58.	*Echinochloa colona* (L.) Link		✓
59.	*Eleusine coracana* (L.) Gaertn.		✓
60.	*Eleusine indica* (L.) Gaertn.		✓
61.	*Eragrostiella bifaria* (Vahl) Bor		✓
62.	*Eragrostis atrovirens* (Desf.) Trin. ex Steud.		✓

Sl. No.	**Name of the Plants**	**Plants growing in Water lodged area**	**Marginal Plants**
63.	*Eragrostis ciliata* (Roxb.) Nees		✓
64.	*Eragrostis gangetica* (Roxb.) Steud.		✓
65.	*Eragrostis nutans* (Retz.) Nees ex Steud.		✓
66.	*Eragrostis tenella* (L.) P. Beauv. ex Roeus. & Schult.		✓
67.	*Eragrostis tremula* (Lam.) Hochst. ex Steud.		✓
68.	*Eragrostis unioloides* (Retz.) Nees ex Steud.		✓
69.	*Eriochloa procera* (Retz.) C.E. Hubb.		✓
70.	*Hackelochloa granularis* (L.) O. Kuntze		✓
71.	*Heteropogon contortus* (L.) P. Beauv. ex Roem & Schult.		✓
72.	*Hordeum vulgare* L.		✓
73.	*Imperata cylindrica* (L.) Beauv.		✓
74.	*Ischaemum indicum* (Houtt.) Merr.		✓
75.	*Iseilema anthephoroides* Hack.		✓
76.	*Iseilema prostratum* (L.) Anderss.		✓
77.	*Leersia hexandra* Sw.		✓
78.	*Leptochloa panicea* (Retz.) Ohwi.		✓
79.	*Oplismenus burmanii* (Retz.) P. Beauv.		✓
80.	*Oryza rufipogon* Griff.	✓	
81.	*Oryza sativa* L.	✓	
82.	*Panicum paludosum* Roxb.	✓	

Sl. No.	**Name of the Plants**	**Plants growing in Water lodged area**	**Marginal Plants**
83.	*Panicum psilopodium* Trin.	✓	
84.	*Panicum repens* L.	✓	
85.	*Panicum trypheron* Schult.		✓
86.	*Paspalidium flavidum* (Retz.) A. Camus		✓
87.	*Paspalidium geminatum* (Forssk.) Stapf		✓
88.	*Paspalum distichum* L.		✓
89.	*Paspalum scrobiculatum* L.		✓
90.	*Pennisetum pedicellatum* Trin.		✓
91.	*Perotis indica* (L.) O. Kuntze		✓
92.	*Phalaris minor* Retz.		✓
93.	*Polypogon monspeliensis* (L.) Desf.		✓
94.	*Rottboellia cochinchinensis* (Lour.) W.D. Claylon		✓
95.	*Saccharum bengalense* Retz.		✓
96.	*Saccharum officinarum* L.		✓
97.	*Saccharum spontaneum* L.		✓
98.	*Sacciolepis interrupta* (Willd.) Stapf	✓	
99.	*Sacciolepis myosuroides* (R. Br.) A. Camus		✓
100.	*Sehima nervosum* (Rottb.) Stapf		✓
101.	*Setaria glauca* (L.) P. Beauv.		✓
102.	*Setaria intermedia* Roem. & Schult.		✓
103.	*Setaria pumila* (Poir.) Roem. & Schult.		✓

Sl. No.	Name of the Plants	Plants growing in Water lodged area	Marginal Plants
104.	*Setaria verticillata* (L.) P. Beauv.		✓
105.	*Sporobolus diander* (Retz.) P. Beauv.		✓
106.	*Themeda quadrivalvis* (L.) O. Kuntze		✓
107.	*Thysanolaena maxima* (Roxb.) Kuntze		✓
108.	*Triticum aestivum* L.	✓	
109.	*Vetiveria zizanioides* (L.) Nash	✓	
110.	*Zea mays* L.		✓

Table - 27: Sedges and Grasses which are very common in Gautam Buddha Wildlife Sanctuary, Hazaribag

Sl. No.	Name of the Plants
1.	*Bulbostylis barbata* (Rottb.) Kunth ex Clark
2.	*Cyperus difformis* L.
3.	*Cyperus iria* L.
4.	*Cyperus rotundus* L.
5.	*Mariscus paniceus* (Rottb.) Vahl.
6.	*Schoenoplectus articulatus* (L.) Palla
7.	*Schoenoplectus supinus* (L.) Palla
8.	*Alloteropsis cimicina* (L.) Stapf.
9.	*Aristida adscensionis* L.
10.	*Aristida setacea* Retz.
11.	*Bambusa arundinacea* (Retz.) Willd.
12.	*Bothriochloa pertusa* (L.) A. Camus
13.	*Brachiaria ramosa* (L.)
14.	*Chloris barbata* Sw.
15.	*Chrysopogon aciculatus* (Retz.) Trin.
16.	*Cynodon dactylon* (L.) Pers.
17.	*Dactyloctenium aegyptium* (L.) Beauv.
18.	*Dendrocalamus strictus* (Roxb.) Nees
19.	*Desmostachya bipinnata* (L.) Stapf
20.	*Dichanthium annulatum* (Forssk.) Stapf
21.	*Digitaria ciliaris* (Retz.) Koeler
22.	*Echinochloa colona* (L.) Link
23.	*Eleusine indica* (L.) Gaertn.

Sl. No.	Name of the Plants
24.	*Eragrostis tenella* (L.) P. Beauv. ex Roem and Schult.
25.	*Eragrostis unioloides* (Retz.) Nees ex steud.
26.	*Heteropogon contortus* (L.) P. Beauv ex Roem and Schult.
27.	*Oplismenus burmanii* (Retz.) P. Beauv.
28.	*Oryza rufipogon* Griff.
29.	*Oryza sativa* L.
30.	*Panicum repens* L.
31.	*Paspalidium flavidum* (Retz.) A. Camus
32.	*Paspalum scrobiculatum*
33.	*Pennisetum pedicellatum* Trin.
34.	*Saccharum officinarum* L.
35.	*Saccharum spontaneum* L.
36.	*Setaria glauca* (L.).
37.	*Sporobolus diander* (Retz.) P. Beauv.
38.	*Triticum aestivum* L.
39.	*Zea mays* L.

Table - 28: Sedges and Grasses used for different economic activities

Sl. No.	Name of Plants	Uses
1.	*Bulbostylis barbata* (Rottb.) Kunth ex Clark	Used in Dysentery
2.	*Cyperus iria* L.	Stems are woven into mats; also used as astringent, stomachic and tonic.
3.	*Cyperus rotundus* L.	Roots are used in making perfumes and in stomach and bowel complaint
4.	*Aristida setacea* Retz.	Culms are used in making broom, brushes, screens and frames for paper manufacture.
5.	*Bambusa arundinacea* (Retz.) Willd.	Culms are used in construction purposes. These also yield a good quality paper. Young shoots are pickled or made into curries.
6.	*Bothriochloa bladhii* (Retz.) S.T. Blake	Culms yield a good paper.
7.	*Cymbopogon martinii* (Roxb.) Wats.	Source of Palmarosa oil, used in soap and cosmetics. It is also used for lumbago and stiff joints.
8.	*Cynodon dactylon* (L.) Pers.	Leaves are very auspicious and used extensively in all religious festivals. The local people employ the juice of the leaves in healing the cuts.
9.	*Dendrocalamus strictus* (Roxb.) Nees.	Culms are employed for rafters battens, baskets, sticks, furniture fishing rods, etc; also used for paper pulp.
10.	*Desmostachya bipinnata* (L.) Stapf	Often used in Hindu ceremonies.
11.	*Eleusine coracana* (L.) Gaertn.	It's grains are used in cakes puddlings and in preparation of alcoholic beverage; also useful in biliousness.
12.	*Eleusine indica* (L.) Gaertn.	Culms are used for making hats.

Sl. No.	Name of Plants	Uses
13.	*Eragrostis tremula* (Lans.) Hochst ex steud.	Grains are eaten in time of scarcity.
14.	*Eragrostis unioloides* (Retz.) Nees ex steud.	Used as green manure.
15.	*Heteropogon contortus* (L.) P. Beauv. ex Roem and Schult.	Used in manufacture of paper. Roots are used as stimulant and diuretic.
16.	*Hordeum vulgare* L.	Grains are used in the form of aata and sattu. It is also used as coolent.
17.	*Imperata cylindrica* (L.) P. Beauv.	Plant is used in manufacture of paper. Culms are used in making ropes, brushes, mats, etc. It is also employed as packing material.
18.	*Oryza rufipogon* Griff.	Grains are used in scarcity.
19.	*Oryza sativa* L.	Rice, the main food for Indian comes from this.
20.	*Panicum paludosum* Roxb.	Grains are used by tribes for making Cake like preparation.
21.	*Panicum psilopodium* Trin.	Grains are used in the preparation of alcoholic beverages.
22.	*Panicum repens* L.	Used for turfs and lawns.
23.	*Panicum trypheron* Schult.	It's grains are used in making breeds in the time of scarcity.
24.	*Phalaris minor* Retz.	Grains are used as birdfeed.
25.	*Polypogon monspeliensis* (L.) Desf.	Sometimes it is cultivated in gardens for its beautiful panicle.
26.	*Rottboellia cochinchinesis* (Lour.) W.D. Clayton	Used in making mats.
27.	*Saccharum bengalense* Retz.	Fibres of the leaf sheaths are used for making mats and ropes.
28.	*Saccharum Officinarum*	Used in making sugar and 'Gud'.
29.	*Saccharum spontaneum* L.	Plant yields a good paper pulp.

Sl. No.	Name of Plants	Uses
30.	*Triticum aestivum* L.	Most important food obtained from this in the form of Ata, Maida, Suji etc.
31.	*Vetiveria zizanioides* (L.) Nash	'Khas' obtains from this.
32.	*Zea mays* L.	Grains are used in the form of popcorn, sattu, etc.

Table - 29: Showing Plants which are used as fodders

Sl. No.	Name of Plants	Used as green fodder	Used as dried fodder
1.	*Cyperus halpan* L.	✓	
2.	*Cyperus iria* L.	✓	✓
3.	*Cyperus rotundus* L.	✓	
4.	*Mariscus paniceus* (Rottb.) Vahl	✓	
5.	*Apluda mutica* L.	✓	
6.	*Aristida setacea* Retz.	✓	
7.	*Bambusa arundinacea* (Retz.) Willd.	✓	
8.	*Bothriochloa bladhii* (Retz.) S.T. Blake	✓	
9.	*Brachiaria ramosa* (L.) Stapf	✓	
10.	*Chloris barbata* Sw.	✓	
11.	*Cynodon dactylon* (L.) Pers.	✓	
12.	*Dactyloctenium aegyptium* (L.) Beauv.	✓	
13.	*Dendrocalamus strictus* (Roxb.) Nees	✓	
14.	*Desmostachya bipinnata* (L.) stapf	✓	
15.	*Digitaria bicornis* (Lans.) Roem. and Schult. ex Loud.	✓	
16.	*Digitaria ciliaris* (Retz.) Koeler	✓	
17.	*Digitaria longiflora* (Retz.) Pers.	✓	
18.	*Echinochloa colona* (L.) Link	✓	
19.	*Eleusine indica* (L.) Gaertn.	✓	
20.	*Eragrostiella bifaria* (Vahl) Bor	✓	

Sl. No.	Name of Plants	Used as green fodder	Used as dried fodder
21.	*Eragrostis atrovirens* (Desf.) Trin ex Steud.	✓	
22.	*Eragrostis ciliata* (Roxb.) Steud.	✓	
23.	*Eragrostis gangetica* (Roxb.) Steud.	✓	
24.	*Eragrostis tenella* (L.) P. Beauv.	✓	
25.	*Eragrostis tremula* (Lans.) Hochst. ex Steud.	✓	
26.	*Eragrostis unioloides* (Retz.) Nees ex steud.	✓	
27.	*Hordeum vulgare* L.	✓	✓
28.	*Ischaemum indicum* (Houtt.) Merr	✓	
29.	*Iseilema prostratum* (L.) Anderss.	✓	
30.	*Leersia hexandra* Sw.	✓	✓
31.	*Oryza rufipogon* Griff.	✓	
32.	*Oryza sativa* L.		✓
33.	*Panicum paludosum* Roxb.	✓	✓
34.	*Panicum psilopodium* Trin.	✓	
35.	*Panicum repens* L.	✓	
36.	*Panicum trypheron Schult.*	✓	
37.	*Paspalidium flavidum* (Retz.) A.Camus	✓	
38.	*Paspalum distichum* L.	✓	
39.	*Paspalum scrobiculatum* L.	✓	
40.	*Pennisetum pedicellatum* Trin.	✓	✓
41.	*Perotis indica* (L.) O.Kuntze	✓	

Sl. No.	Name of Plants	Used as green fodder	Used as dried fodder
42.	*Phalaris minor* Retz.	✓	
43.	*Polypogon monspeliensis* (L.) Desf.	✓	
44.	*Rottboellia cochinchinensis* (Lour.) W.D. Clayton	✓	
45.	*Saccharum officinarum*	✓	
46.	*Sacciolepis myosuroides* (R. Br.) A. Camus	✓	
47.	*Sehima nervosum* (Rottb.) Stapf	✓	
48.	*Setaria glauca* (L.) P. Beauv.	✓	
49.	*Setaria intermedia* Roem. and Schult.	✓	
50.	*Thysanolaena maxima* (Roxb.) kuntze	✓	
51.	*Triticum aestivum* L.		✓
52.	*Zea mays* L.	✓	

Table - 30: Sedges and Grasses considered as weeds of farms and gardens

Sl. No.	Name of the Plants	Gregarious	Frequent	Occasional
1.	*Cyperus castaneus* Willd.		✓	
2.	*Cyperus difformis* L.	✓		
3.	*Cyperus distans* L.			✓
4.	*Cyperus dubius* Rottb.			✓
5.	*Cyperus halpan* L.			✓
6.	*Cyperus iria* L.	✓		
7.	*Cyperus procerus* Rottb.		✓	
8.	*Cyperus rotundus* L.	✓		
9.	*Cyperus tenuispica* Steud.		✓	
10.	*Fimbristylis alboviridis* C.B. Clarke	✓		
11.	*Fimbristylis argentea* (Rottb.) Vahl	✓		
12.	*Fimbristylis miliacea* (L.) Vahl.		✓	
13.	*Fimbristylis tetragona* R.Br.		✓	
14.	*Fuirena ciliaris* (L.) Roxb.		✓	
15.	*Kyllinga brevifolia* Rottb.	✓		
16.	*Kyllinga bulbosa* Beauv.	✓		
17.	*Mariscus paniceus* (Rottb.)		✓	
18.	*Pycreus flavidus* (Retz.) Koyama	✓		
19.	*Pycreus pumilus* (L.) Nees ex Clarke	✓		
20.	*Schoenoplectus supinus* (L.) palla	✓		
21.	*Alloteropsis cimicina* (L.) stapf		✓	

Sl. No.	Name of the Plants	Gregarious	Frequent	Occasional
22.	*Brachiaria ramosa* (L.) Stapf	✓		
23.	*Dactyloctenium aegyptium* (L.) Beauv.	✓		
24.	*Digitaria ciliaris* (Retz.) Koeler	✓		
25.	*Digitaria longiflora* (Retz.) Pers.			✓
26.	*Echinochloa colona* (L.) Link	✓		
27.	*Eleusine indica* (L.) Gaertn.	✓		
28.	*Eragrostiella bifaria* (Vahl) Bor		✓	
29.	*Eragrostis unioloides* (Retz.) Nees ex Steud.		✓	
30.	*Eriochloa procera* (Retz.) C.E. Hubb.			✓
31.	*Ischaemum indicum* (Houtt.) Merr.			✓
32.	*Oplismenus burmanii* (Retz.) P.Beauv.	✓		
33.	*Oryza rufipogon* Griff.		✓	
34.	*Panicum psilopodium* Trin.		✓	
35.	*Panicum repens* L.	✓		
36.	*Paspalidium flavidum* (Retz.) A. Camus	✓		
37.	*Paspalum distichum* L.		✓	
38.	*Paspalum scrobiculatum* L.		✓	
39.	*Phalaris minor* Retz.			✓
40.	*Polypogon monspeliensis* (L.) Desf.			✓

Sl. No.	Name of the Plants	Gregarious	Frequent	Occasional
41.	*Sacciolepis interrupta* (Willd.) Stapf			✓
42.	*Setaria glauca* (L.) P. Beauv.	✓		
43.	*Setaria intermedia* Roem. and Schult.		✓	
44.	*Setaria pumila* (Poir.) Roem. and Schult.		✓	

Table - 31: Showing name of the Sedges and Grasses which are uncommon in Gautam Buddha Wildlife Sanctuary

Sl. No.	Name of the Plant
1.	*Carex cruciata* Wahlenb.
2.	*Cyperus bulbosus* Vahl.
3.	*Cyperus dubius* Rottb.
4.	*Cyperus nutans* Vahl.
5.	*Cyperus Pangorei* Rottb.
6.	*Cyperus polystachyos* Rottb.
7.	*Cyperus pygmaeus* Rottb.
8.	*Fimbristylis polytrichoides* (Retz.) R. Br.
9.	*Apluda mutica* L.
10.	*Chloris dolichostachya* Lagasca.
11.	*Coix lachryma-jobi* L.
12.	*Cymbopogon martinii* (Roxb.) Wats
13.	*Dichanthium ischaemum* (L.) Roberty
14.	*Digitaria longiflora* (Retz.) Pers.
15.	*Eragrostiella bifaria* (Vahl) Bor.
16.	*Eragrostis ciliata* (Roxb.) steud.
17.	*Eriochloa procera (Retz.) C.E. Hubb.*
18.	*Hackelochloa granularis* (L.) O. Kuntze.
19.	*Iseilema anthephoroides* Hack.
20.	*Leersia hexandra* Sw.
21.	*Leptochloa panicea* (retz.) Ohwi.
22.	*Panicum trypheroan* Schult.
23.	*Paspalidium geminatum* (Forssk.) Stapf
24.	*Rottboellia cochinchinensis* (Lour.) W.D. Clayton

Sl. No.	**Name of the Plant**
25.	*Sacciolepis interrupta* (Willd.) Stapf.
26.	*Sacciolepis myosuroides* (R.Br.) A. Camus.
27.	*Sehima nervosum* (Rottb.) Stapf.
28.	*Setaria verticillata* (L.) P. Beauv.

Table - 32: Artificial key to the Grasses found in Gautam Buddha Wildlife Sanctuary

Characters specific to groups									
Tall, arborescent and stout	Grasses have fruits as white nuts	Grasses when fruits are not a white nut							
		Grasses with trifid awns	Grasses without trifid awns						
			Inflorescence a plumose panicle	Inflorescence/ a single raceme/spike/ cylindrical	Inflorescence terminal whorled, digitate	Inflorescence/ of linear racemes arranged on a single long or short central axis	Inflorescence with bare branches or pediceled spikelets	Inflorescence with spathes in groups of 2	Inflorescence with spathes in heads
A	**B**	**C**	**D**	**E**	**F**	**G**	**H**	**I**	**J**
Bambusa arundinacea	*Coix lachryma-jobi*	*Aristida adscensionis*	*Phalaris minor*	*Heteropogon contortus*	*Alloteropsis cimicina*	*Brachiaria ramosa*	*Chrysopogon aciculatus*	*Cymbopogon martinii*	*Apluda mutica*
Dendrocalamus strictus		*Aristida setacea*	*Polypogon monspeliensis*	*Imperata cylindrica*	*Bothriochloa bladhii*	*Brachiaria reptans*	*Eragrostiella bifaria*		*Iseilema anthephoroides*
			Saccharum bengalense	*Oryza rufipogon*	*Bothriochloa pertusa*	*Desmostachya bipinnata*	*Eragrostis atrovirens*		*Iseilema prostratum*
			Saccharum spontaneum	*Pennisetum pedicellatum*	*Chloris barbata*	*Digitaria bicornis*	*Eragrostis ciliata*		*Themeda quadrivalvis*
			Thysanolaena maxima	*Perotis indica*	*Chloris dolichostachya*	*Digitaria ciliaris*	*Eragrostis gangetica*		
				Phalaris minor	*Cynodon dactylon*	*Digitaria longiflora*	*Eragrostis nutans*		
				Polypogon monspeliensis	*Dactyloctenium aegyptium*	*Echinochloa colona*	*Eragrostis tenella*		

Characters specific to groups									
Tall, arborescent and stout	**Grasses have fruits as white nuts**	**Grasses when fruits are not a white nut**							
		Grasses with trifid awns	**Grasses without trifid awns**						
			Inflorescence a plumose panicle	**Inflorescence/ a single raceme/spike/ cylindrical**	**Inflorescence terminal whorled, digitate**	**Inflorescence/ of linear racemes arranged on a single long or short central axis**	**Inflorescence with bare branches or pediceled spikelets**	**Inflorescence with spathes in groups of 2**	**Inflorescence with spathes in heads**
				Rottboellia cochinchinensis	*Dicanthium annulatum*	*Hackelochloa granularis*	*Eragrostis tremula*		
				Sacciolepis interrupta	*Dicanthium ischaemum*	*Oplismenus burmanii*	*Eragrostis unioloides*		
				Sacciolepis myosuroides	*Eleusine indica*	*Paspalidium flavidum*	*Leersia hexandra*		
				Sehima nervosum	*Ischaemum indicum*	*Paspalidium geminatum*	*Leptochloa panicea*		
				Setaria glauca	*Paspalum distichum*	*Vetiveria zizanioides*	*Panicum paludosum*		
				Setaria intermedia	*Paspalum scrobiculatum*		*Panicum psilopodium*		
				Setaria pumila			*Panicum repens*		
				Setaria verticillata			*Panicum trypheron*		
							Sporobolus diander		
				Cultivable					

Characters specific to groups									
Tall, arborescent and stout	Grasses have fruits as white nuts	Grasses when fruits are not a white nut							
		Grasses with trifid awns	Grasses without trifid awns						
			Inflorescence a plumose panicle	Inflorescence/ a single raceme/spike/ cylindrical	Inflorescence terminal whorled, digitate	Inflorescence/ of linear racemes arranged on a single long or short central axis	Inflorescence with bare branches or pediceled spikelets	Inflorescence with spathes in groups of 2	Inflorescence with spathes in heads
				Eleusine coracana					
				Hordeum vulgare					
				Oryza sativa					
				Saccharum officinarum					
				Triticum aestivum					
				Zea mays					